P9-BIY-133

ULTRASONICS IN PACKAGING AND PLASTICS FABRICATION

ULTRASONICS IN PACKAGING AND PLASTICS FABRICATION

Ralph H. Thomas, Sr.
Director of Package Concept Development
Bristol-Myers Products Division
Bristol-Myers Company

Cahners Books
A Division of Cahners Publishing Company, Inc.
89 Franklin Street, Boston, Massachusetts 02110

Library of Congress Cataloging in Publication Data

Thomas, Ralph H 1929–
Ultrasonics in packaging and plastics fabrication.

1. Ultrasonic welding. 2. Plastics—Welding.
3. Plastics in packaging. I. Title.
TS228.92.T47 668.4'15 73-76443
ISBN 0-8436-1102-2

International Standard Book Number: 0-8436-1102-2
Library of Congress Catalog Card Number: 73-76443

Printed in the United States of America.

This book is dedicated to those individuals who have pioneered the packaging research field and to those devoted to furthering the academic knowledge of this specialized endeavor.

Table of Contents

Foreword

It has long been recognized that there existed a need for a comprehensive and modern treatise on packaging materials and techniques employed in the cosmetic industry. This has become increasingly important in the past decade because of the wider selection of basic packaging materials which have become available, improved methods for the design and development of containers, and the proliferation of products which require distinctive and specially designed containers for enhanced marketability and prolonged stability.

In addition to technological advances there exists the ever-increasing demands of regulatory agencies concerned with the safety and utility of products on the American market. A packaging engineer must be as completely aware of the regulatory agency controls as he is of the properties and applications of the basic tools of his profession.

This book is authored by an expert in the packaging research field, and will serve to fill the need for a comprehensive reference in modern packaging research and development.

The College of Pharmaceutical Sciences, Columbia University, is pleased to present this volume as part of its programs of service to the cosmetic industry. The book was originally written as a textbook for students enrolled in the Cosmetic Packaging Technology course in the College's evening extension program in Cosmetic Sciences.

Joseph L. Kanig, Dean
College of Pharmaceutical Sciences
Columbia University

Preface

Although the technology of ultrasonics extends back to the early years of the twentieth century, the industry that has developed from this technology is but a few years old. Historically speaking, it is an industry still in its infancy, yet one that has made scientific and commercial progress at a challenging pace. Literally thousands of applications in consumer packaging and other products have been manufactured and sold in which ultrasonics has made a vital contribution to their success.

As a new industry has risen out of the scientific developments that have made its start possible, advances in its technology have continued. These advances in ultrasonics have been particularly striking in plastics fabricating, covering all fields of plastics usage from consumer packaging to sewing. There have been countless special techniques developed for the processing and assembly of all types of plastics products. This progress is reflected in the activities of scientific and trade associations, with equipment manufacturers and process developers, and in professional publications as well as in bulletins and other literature published by the companies active in the field.

As the industry grows and the scientific literature multiplies, the need for a compendium that gathers this information into a single volume becomes apparent. This is especially the case with the application of ultrasonics in packaging. For twenty years, industrial and technological developments related to packaging have been so rapid that it is difficult to keep abreast of the changes and improvements that have occurred.

As an individual who was privileged to be active as a packaging engineer in the early stages of ultrasonics applications in the fabrication of display promotions, I have been a witness and participant in this phenomenal growth. As the technology developed, it became clear that a manuscript summarizing these developments would be an asset to progress as well as an historic record.

It is the author's desire to impart through this book broad knowledge of the field of ultrasonic sciences as applied to the field of consumer products.

Many people participated in making possible this collective volume, many more than those named in the references. Above all, thanks must be expressed to the organizations that research and manufacture ultrasonic equipment.

In addition, my personal thanks go to the entire industry for their participation in these great technological and industrial developments. Among others too numerous to mention, I should like to thank and credit Lowndes Rourke of Branson Ultrasonics, Inc.; C. J. Williams, Consultant; and Dr. L. Balamuth, consultant to Ultrasonics Systems, Inc., whose personal assistance made it possible for this book to materialize.

I would further like to credit the American Welding Society, both its members and staff, who have given such able guidance and help to the entire industry.

The future of ultrasonics, both as a science and as an industry, is undoubtedly as promising as its recent past, and it is hoped that just as this book seeks to be a summary of developments of the past twenty years, it may serve to help catalyze developments for many years ahead.

Ralph H. Thomas

Section I
Theory of Ultrasonics

Chapter 1
Introduction and Prospects

The markets for consumer packaged goods have expanded to include many items once considered to be luxuries. The increased sales of many products that are now classified as everyday commodities have occurred through the continuing efforts of manufacturers to meet consumer needs. The consumer can now obtain many products in a variety of forms. This is often accomplished through special packaging. A typical example would be deodorants and antiperspirants that are available in liquid, lotion roll-on, aerosol pressure packs, pads, plastic squeeze bottles, sponge applicators, stick form, gelatins, creams, brush applicators and pills.

Ultrasonics first started to make major contributions to the high-speed bonding and fusing of plastics packaging component parts about ten years ago. Previously, ultrasonics had only been used in packaging for cleaning, filling, mixing, sealing and heating. Starting in the early 1960s there were many new concepts in packaging designs that demanded more positive and assured securement than strictly undercuts, friction and cementing of plastic component parts. Fortunately, the materials and techniques to meet these needs were available. The advent of new plastic resins and container fabrication techniques, including the field of ultrasonic sciences, have been responsible for significant advancement of new packaging conveniences and product applications.

This book details and defines the science of ultrasonics, from theory and equipment to applications. Its intention is to provide the packaging engineer with a basic understanding of the known principles of ultrasonics as they apply to packaging needs. Specific examples of new equipment and their application to commercial practice are provided to illustrate the spectacular developments that have occurred in the past ten years. For example, ultrasonic motors have been reduced from the size of a standard one-quarter h.p. down almost to the size of a fountain pen, thus providing many more areas of commercial applications.

Developments in ultrasonics that relate directly to packaging applications are numerous. For example, new features of ultrasonic sewing machines along with plastic film skin and shrink packaging by ultrasonics have greatly enhanced new and potential applications. Recently an ultrasonic injection machine for plastics was introduced at the Plastics Exposition in Dusseldorf, Germany. This machine requires no screw, piston, heating cylinder, blender, heating jacket or pressure ram at all. With a capacity to produce shots above 5 ounces, it requires no more floor space than a large-size office desk. Various research projects of major proportions have been underway by several of the ultrasonics equipment and process manufacturers in all related fields of packaging, from attempting to re-orient the molecular structure of a molded plastics polyolefin to the capability of welding under pressure the 1-in. mounting cup of aerosol valves to the metal can body.

The Ultrasonics Industry

Today, an industry utilizing ultrasonic principles that nature gave to bats eons ago is coming into its own. The first caveman to try to express himself in something more sophisticated than a grunt became a true pioneer of this industry by raising his voice and launching us on our way up the electro-magnetic frequency scale. Audible sound is at the bottom of this scale, which leads directly to the soprano C above high C that can shatter glass, and from there to the even higher-pitched whistle that only dogs can hear. The same scale keeps moving upward from the dog whistles into radio waves, and from these still higher on the vibratory scale until we get to waves which oscillate so rapidly that they carry heat. From there on up, the heat is intensified to the point where visibility begins at the red end of the spectrum and continues through to violet. Above violet we get ultra-violet, which burns our skin; and higher still until we run into x-rays, which can be fatal. For the ultrasonics industry, the most useful vibrations exist just below radio waves; and this is the area in which rapid-fire developments in industry, research and medicine are making ultrasonics one of today's fastest-growing technologies.

Applications for this silent form of energy are fantastic. The first immediately apparent applications are in cleaning, where bombarding a surface with billions of ultrasonic bubbles produces a cleanness unmatched by any other known method. Computer manufacturers are, for example, currently using automatic ultrasonic cleaning systems for removing microscopic dirt particles from disks in computer memories. Automobile manufacturers use ultrasonic baths to clean automobile transmission parts during assembly. And ultrasonic toothbrushes and dishwashers are already on the drawing boards. Future applications seem almost boundless, and are limited only by imagination and cost.

Total ultrasonics industry sales in 1969 were estimated at $65-$70 million, based on a study conducted by *Dun's Review*. Projections show the industry growing to $162 million by 1973. The industry consists of two major markets: commercial/industrial and medical/dental with the consumer market rapidly emerging as a third major market.

Commercial/Industrial Market. Industry volume in 1970 is rising at a 15%–20% annual rate, and is expected to accelerate in the next decade. Commercial equipment sales in 1969 reached an estimated $55-$60 million, while projections for 1973 are that sales are expected to climb to $122 million.

By far the largest application, accounting for about $21-$23 million in 1969, was ultrasonic cleaning. Other growth areas are instrumentation, assembly, electronics, packaging and textiles. In the industrial field, some of the applications are welding, impact grinding, cleaning, and the technique for embedding inserts in plastics.

Medical Market. Medical ultrasonics is a $6-$7 million market at present, and industry sources estimate the sale of this equipment will soar to $30 million by 1973. New concepts are being developed, such as an ultrasonic suture device, an ultrasonic vial and equipment for removing tissue by micro-chopping of biological organisms.

Consumer Market. Industry experts project a $10 million consumer market in 1973. New products, such as an ultrasonic home cleaner, an ultrasonic razor, an ultrasonic water pick, an ultrasonic toothbrush, an ultrasonic hobby kit and an ultrasonic engraving unit are being developed with small ultrasonic motors.

Background for Commercial Applications

The basic ultrasonic energy is a mechanical energy in the form of vibrations, normally above 18,000 cycles per second, which is harnessed to perform many thermal and non-thermal reactions, either exothermic or non-exothermic, depending on the combination requirements. This relationship of plastic materials as to the nature of their structure and compatibility is important to efficient performance in ultrasonic application. Of all known technical data about bonding plastics, ultrasonics is capable of fusing the widest range of compatible and noncompatible films and sheets.

Ultrasonic energy, which was first used practically for underwater submarine detection during World War I, has come a long way. The packaging industry today is applying ultrasonics in over two thousand methods with more efficient products and packages at a lower cost, and with more consumer function and convenience application. The first experiments in using ultrasonics to join plastic film material were conceived in the early 1940s. Since that time, equipment and technology have been pioneered to develop the principle of bonding to the point where rigid sheet, film and molded parts of plastic can now be welded ultrasonically.

Insulation techniques have now been developed wherein over a given seal area during the same cycle, an entire section can have interrupted weld seals as required, and/or other sub-assemblies as connections to the overall finished unit.

High-speed, practical and efficient ultrasonic assembly of plastic components is basically the result of assuring various prerequisites. The single or multiple effects of the following factors will provide the final results:

1. Joint or interface design
2. Material
3. Shape and size of part
4. Wall and cross-section thickness
5. Near or far field
6. Shape of horn
7. Amplitude of horn
8. Alignment
9. Nesting
10. Cleaning and preparation of surface
11. Ambient temperature of components
12. Energy delivered to the part
13. Compatibility of plastics to be joined
14. Internal or external dampening elements
15. Size of weld joint requirements

Today the ultrasonic welding of plastics and metals has provided an expedient technique of positive bonding which is applicable to high-speed production requirements. This technique is now available for metals, plastics, combinations, as well as insert-welding metals to plastics.

The most often used injection molded materials can be ultrasonically welded without the use of solvents or adhesives. Weldability of these materials depends on the melting temperature, modules of elasticity, impact resistance, co-efficient of friction, design and shape, and thermal conductivity. In general, the softer the plastic, the more difficult it is to control, compress and weld. Low modulus materials such as polyethylene and polypropylene can often be welded together, provided the ultrasonic welding horn can be positioned close to the joint areas.

The basic definition and general description of ultrasonic welding is varied. The following description covers the critical concepts of the process:

Ultrasonic welding is a process for joining similar and dissimilar metals or plastics by the introduction of high-frequency vibratory energy into the overlapping metals or plastics in the area to be abutted.

No fluxes or fillers are used, no electrical current passes through the weld material, and usually no heat is applied. The workpieces are clamped together under moderately low static force, and ultrasonic energy is transmitted into the intended weld area. A strong ultrasonic bond is produced without arc or complete melting of the metal or plastic and without the cast structure associated with melting. There is minor thickness deformation over the actual weld joint.

This joining process has demonstrated its versatility in applicable situations

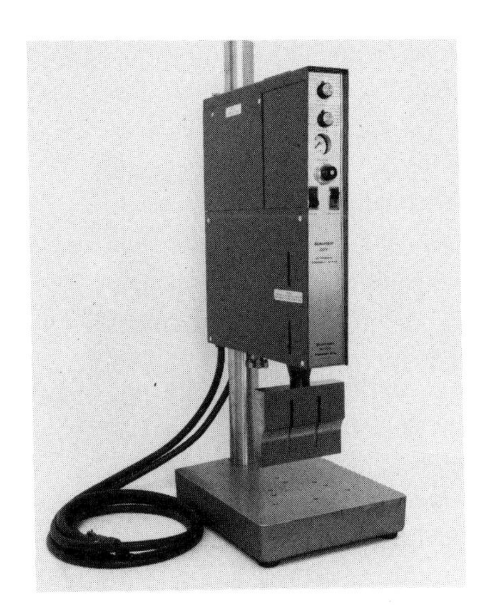

Figure 1.1. Typical ultrasonic.

involving dissimilar-material junctions, and in producing a variety of joint configurations. These principles are used in production in the semi-conductor and micro-circuit industries, within the manufacture of aluminum foils, fabrication of plastic products, metal and plastic combinations and for many applications in the field of packaging. Many other innovations are in the pilot-production stage as equipment is being engineered to meet manufacturing demands. The ultrasonic welding process is utilized by means of spot welding, ring-welding, line-welding, continuous-seam-welding, gang-welding and scan-welding.

In all types of ultrasonic welding, static clamping force is applied approximately normal to the interface between the workpieces. The contacting welding horn oscillates approximately parallel to the plane of this interface.[1] The combined static and oscillating forces cause oscillating stresses at the weld interface, which then result in very local slip between the workpieces in portions of the weld locale, expelling foreign matter and permitting metal-to-metal or plastic-to-plastic contact. Since ultrasonic welds can be made to overlap, it is clear that gross relative motion between the workpieces is not necessary to the process.

The delivered energy converts to heat and a moderate temperature rise occurs, thus altering the properties of the material in the weld zone. Except for this transient alteration in material properties, it appears that heat predominance plays no significant role in the process. It has been established that differences in room-temperature material properties and material thickness cause significant changes in the energy required to make a weld. The relationship of these factors to bonding becomes complicated in calculation, but requires verification with set-up testing prior to firm production application.

Almost every metal and alloy can be ultrasonically welded, although other welding processes may be more economical for certain of the more readily weldable materials. Ultrasonic welding is particularly useful for materials and certain geometries that are difficult or impossible to join by other techniques, for materials that require unusually high power for satisfactory bonding, or for applications that cannot tolerate some of the side effects caused mainly by heat, which may occur with other joining methods.

Actually, various materials differ in weldability according to their composition and material properties. Particular notice should be paid to the relationship between power required, thickness, and hardness.

For a given material and material thickness combination, there are ranges of welding-machine settings of power, clamping force, and weld time that will produce satisfactorily high-strength bonds free from cracks and surface defects. The more difficult to weld materials are considered to be those which require higher power and/or longer weld intervals, or which introduce problems such as short tool tip life and tip sticking.

The ultrasonic weldability of materials is based on known properties of the materials and can be used as guidelines.

Plastics (More details on welding are provided in Chapter 6.)

Polystyrene has excellent acoustical properties; produces strong, smooth joints.

Polycarbonates have high melting temperatures and require high energy levels.

Acetal requires high energy and long ultrasonic exposure because of a low coefficient of friction.

Acrylics are weldable to ABS; applications include: dials, radio cases, and meter housings. In sheet form, joints must be machined.

With polypropylene, horn design is particularly critical; filled compounds need individual testing.

Butyrates, cellulosics and acetates have weldability that varies with formulation and part configuration. These materials perform well in staking and inserting applications (see Chapter 7).

Vinyls as compounds break down chemically during welding; some success has been achieved with rigid vinyls.

Metals

Among the most readily weldable materials are aluminum and its alloys, including high-strength structural alloys. These may be joined in any available form: cast, extruded, rolled, forged, or with any type of heat treatment.

Copper and its alloys (such as brass and gilding metal) as well as other comparatively soft metals are relatively easy to weld. High thermal conductivity of materials does not appear to be a factor in ultrasonic welding as it is in melting-type joining processes.

Satisfactory bonds can be produced in iron and steel of various types, including ingot-iron, low-carbon steels, non-hardening steels such as the austenitic stainless steels, tool and die steels, precipitation-hardening steels, and others. Nickel, titanium, zirconium, and variations of their alloys can also be satisfactorily welded.

No particular difficulties have been encountered with the precious metals:

gold, silver, platinum, or alloys of these metals. These and other metals, such as aluminum, have been satisfactorily bonded to semi-conductors such as germanium and silicon. Good welds are also practical between metal foils or wires and metallized surfaces on glass, ceramic, or silicon, such as printed circuits.

New processes and equipment now have improved welding potential with refractory metals, including molybdenum, columbium, tantalum, and tungsten. Alloys of these metals have demonstrated good metallurgical susceptibility to ultrasonic welding, as have beryllium and rhenium. Thin sheets of these materials have been welded with the use of power-force programming. The quality of the welds appears to depend on the quality of the metal, as well as freedom from contamination and surface or internal defects.

Combining Plastics and Metal

In ultrasonic inserting and encapsulating of metal into plastic, equipment technology can now produce ultrasonic vibrations through the plastic part until they meet at the joining area between metal and plastic. Heat intensity, created by the plastic vibrating against the metal, is sufficient to melt the plastic, permitting the inserts to be embedded into place. In ultrasonic staking, the plastic member is actually deformed and shaped to a desired finish, after mechanically bonding the metal part to the plastic part.

These applications to cosmetic packaging now provide new techniques for construction of compacts, lipstick cases, two-piece cream jars, nailpolish plume assemblies, eyebrow pencils, mascara brushes, eyelash combs, tube colorants, hair curlers, combination combs and many other products.

Note

1. This is the case for metal-to-metal welds; oscillations perpendicular to the surface are for plastic welding.

Chapter 2
Ultrasonic Concepts

Ultrasonics refers to vibrations above the range of human hearing, which is 18,000 cycles per second. Ultrasonic energy is used by industry in a variety of ways, such as for cleaning, thickness gauging, nondestructive flaw-detection, hardness testing, exotic machining, plastics assembly, emulsification, biological cell disruption and many other applications.

How Ultrasonics Works

Standard 115-volt 50 or 60 cycle per second electrical energy is converted to 20,000 cycle per second electrical energy. This high-frequency energy is then fed to a piezoelectric element, called a converter, which transforms the electrical energy to mechanical energy also at 20,000 cycles per second. The resonating converter must be coupled to the object or liquid by means of a mechanical impedance transformer, called a horn. The horn transmits the energy to the part being vibrated or liquid being treated. In welding plastics, for example, the vibrations are transmitted to the joint area where frictional heat is produced to melt the plastic momentarily, causing it to fuse together.[1] When used in a liquid, the vibrations produce intense cavitation (the formation and collapse of small voids at the frequency of 20,000 cycles per second), which disrupts biological cells, making dispersions, and emulsifies immiscible liquids. Contrary to popular science fiction, ultrasonic vibrations cannot be transmitted efficiently through air, so positive coupling with the workpiece or liquid is required.

Plastics Assembly

The ultrasonic plastics assembly unit is an extremely versatile system used to weld plastic to plastic, stake plastic to metal, and insert metal into plastic with the same basic machine, simply by changing horns and adjusting exposure time and air pressure in the cylinders of the actuating mechanism.

Plastics Welding. When high-intensity ultrasonic vibrations are coupled to one piece of a thermoplastic assembly, heat is generated through friction at the interface of the two pieces. Usually within less than a second, the plastic joint melts and becomes solid again to produce a molecular bond.

Designing the joint area and choosing a suitable plastic for ultrasonic welding are two important factors that must be taken into consideration to obtain optimum results.

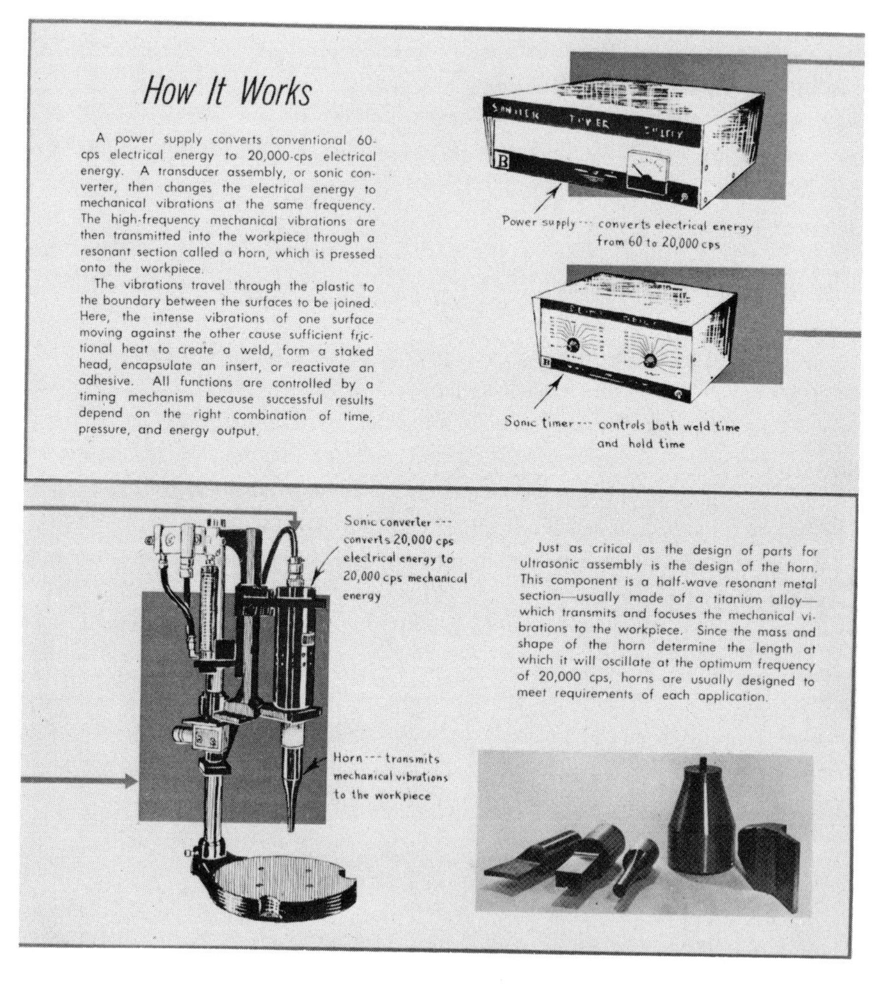

Figure 2.1. Basic ultrasonic process set-up.

Illustrative Applications: Photographic Flashcube. The flashcube, initially designed for ultrasonic welding, is made up of three major parts (see Figure 2.2).

1. Black polystyrene base with four flash bulbs.
2. Foil reflector.
3. Clear polystyrene cover.

The clear cover is slipped over a foil reflector and guided into welding position on the base by four small base-flanges. Welding takes place with the assembled components in an inverted position.

A one-inch-square horn, powered by an ultrasonic power supply, contacts the base of the flashcube to produce a clean weld at high production speeds.

Rotary feed systems have been developed for applications having similar high-production-rate requirements.

Ultrasonic Staking. Metal, as well as other materials, can be joined to plastic by ultrasonically forming a plastic protrusion or stud into a mushroom-shaped

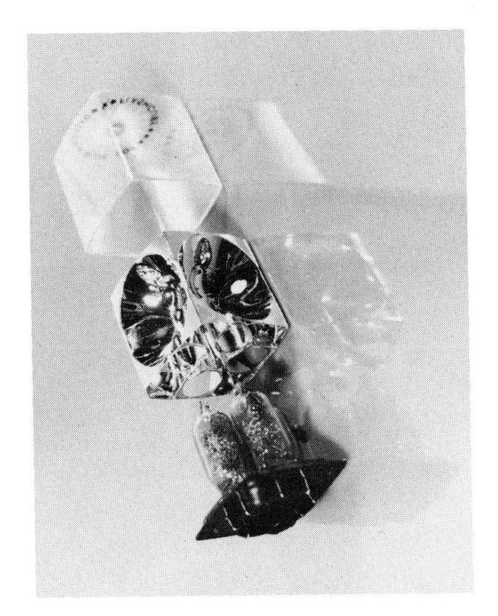

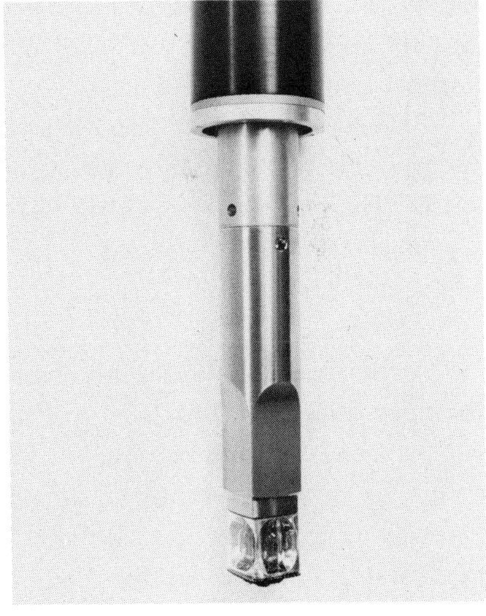

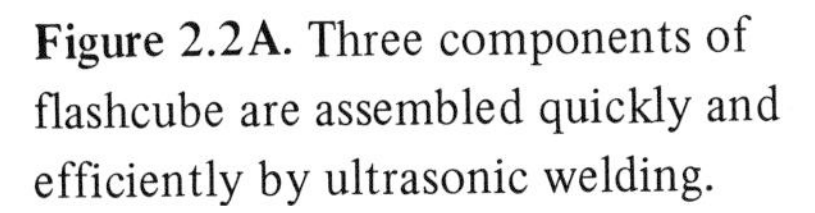

Figure 2.2A. Three components of flashcube are assembled quickly and efficiently by ultrasonic welding.

Figure 2.2B. Flash cube horn design.

head to capture the metal between the head and the plastic base. The results achieved are superior to heat staking; ultrasonic staking is much faster and stronger and produces a more consistent part because the end of the horn, which forms the head, is cool. The problems of inconsistency resulting from temperature variation of a heat-staking tool are eliminated. Also, a clean head is formed because plastic from the formed head will not stick to the cool, ultrasonically activated tool.

Illustrative Application: Breakaway Stud. Ultrasonic staking is a fast and efficient method of assembling metal to plastic (see Figure 2.3). Until now the permanence of the standard stake prevented disassembly for repair without destruction of the plastic part.

With the recent development of the breakaway stud, ultrasonically-staked assemblies can now be disassembled and reassembled with self-tapping screws. This method lowers material costs and saves time in the initial assembly, since no screws are required and several points of assembly can be made simultaneously.

Figure 2.3 shows a cross-section view of a breakaway stud. Rather than molding a standard solid stud, a hole is pre-molded into the stud to accept a self-tapping screw if disassembly is required at a later date. The diameter of the inner hole should be no larger than one-half the outside diameter of the stud, to assure adequate stud strength.

Electrical appliance manufacturers are now using the breakaway stud in the assembly of a small timer mechanism instead of using expensive self-tapping screws. The very small percentage of timers that must be taken apart later for repair are then reassembled with screws. All three points of assembly are staked simultaneously.

Ultrasonic Metal Insertion. Insert molding is a process that has been used for many years to encapsulate metal into plastic. Although the process is slow, cumbersome and quite expensive, it was the only practical technique available before the advent of ultrasonic insertion.

By using high-intensity ultrasonic vibrations, metal can be ultrasonically inserted into a pre-molded hole slightly smaller than the insert. This allows the injection molding machine to be utilized to capacity; less expensive inserts with normal tolerances can be used; mold costs are less and the danger of damaging the mold through improper placement of the insert is eliminated.

Illustrative Application: Eyeglass with Metal Hinge Insert.

Problem: Holding and aligning small metal hinge for insertion into eyeglass frame.

Solution: As illustrated in Figure 2.4, a standard 1/2-in. catenoidal horn with removable tip is used. A cavity in the tip conforms to the shape of the hinge. A small hole, drilled from the center of the cavity and through the horn, connects to a vacuum pump. The hinge is placed in the tip by hand, and the vacuum is sufficient to hold it in place until it is inserted into the frame.

This method obviously speeds production and eliminates error in positioning.

The vacuum horn principle can also be applied to other small parts that are difficult to position.

Figure 2.3. Breakaway plastic stud is staked to metal plate to allow easy disassembly for repair.

Equipment used:

Power Supply: System A
Stand: Basic
Horn: 1/2-in. catenoidal with special tip
Weld time: 0.4 seconds
Hold time: 0.4 seconds

Ultrasonic Machining

A rotary ultrasonic machine tool, which combines the advantages of high-speed rotary cutting tools and reciprocating ultrasonic tools, has recently been developed for high-precision drilling, internal and external thread-forming, and trepanning of hard, brittle materials without the use of an abrasive slurry.

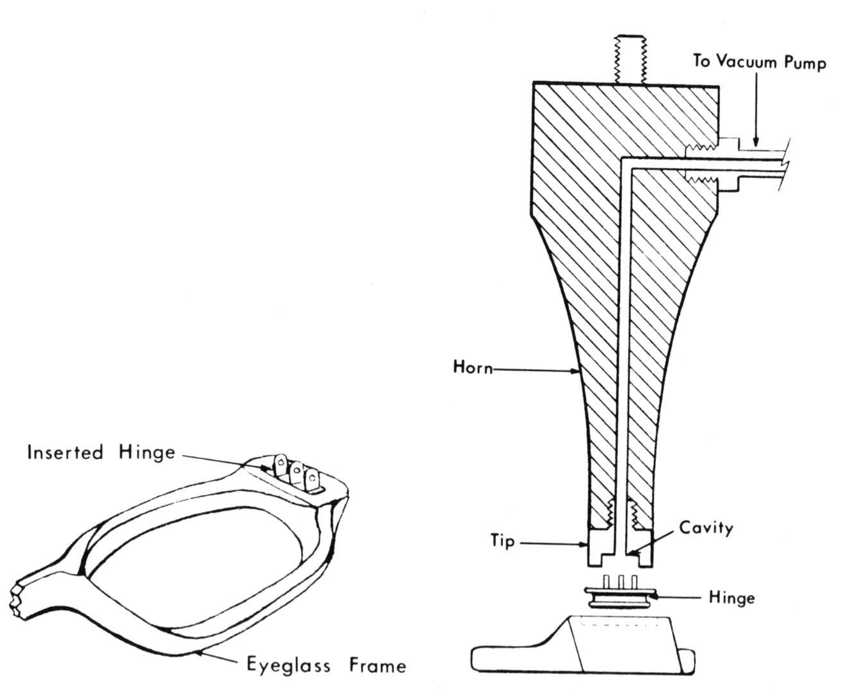

Figure 2.4A. Metal hinge is permanently bonded to plastic eyeglass frame.

Figure 2.4B. End view of eyeglass frame and hinge bonding surface.

The simultaneous rotation and ultrasonic vibration of a diamond abrasive cutting tool reduces friction between the tool and the workpiece, thereby accelerating cutting and extending tool life. This technique requires lower tool pressure, permits the machining of delicate components without cracking, and minimizes shelling at points of entry and breakthrough.

Note

1. The typical application of ultrasonic vibrations to welding plastics is shown in (a). The horn output oscillates perpendicular to the plastic surface.

The typical application of ultrasonic vibrations to welding metals is shown in (b). The horn output oscillates parallel to the plastic surface.

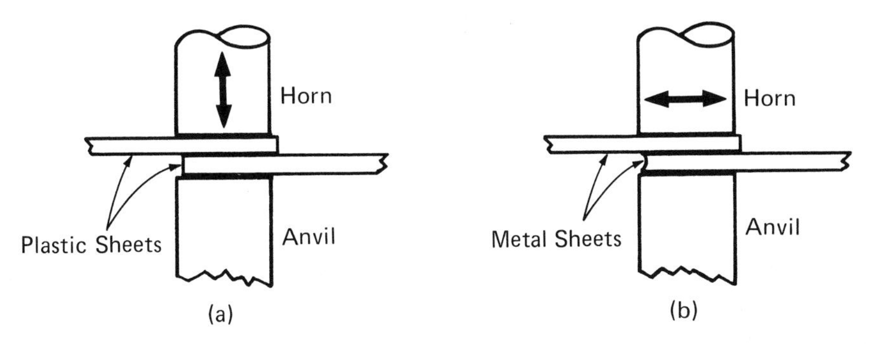

What happens at the interface where welding occurs has not been determined in exact detail to my knowledge to date. It appears quite definite that joining due to melting by heat generation alone is only a part of the story. Some solid state reaction across the boundary apparently occurs due to the coherent (laser-like) acoustic vibrations at the interface. For example, I have welded tungsten to aluminum, where the disparity of melting points is just too great to allow a mutual melting during the weld.

Chapter 3
Biological Applications

The forces generated at the output of the ultrasonic motor or vibrator vary in time in a simple harmonic, periodic fashion. This type of motion is the same as that of a weight bobbing on the end of a vibrating spring. It is in this simple kind of mechanical vibration and motion that we find the keys to understanding the great variety of uses of ultrasonic energy in biological systems.

Every use of ultrasonics starts at the site where the ultrasonic vibrations create the forces which, in turn, act upon the object or area under treatment. It is important, therefore, to know the nature of these forces and their possible effects on the biological system involved.

Let us try to get an amoeba-eye view of an organism in a liquid through which ultrasonic waves are passing. The situation may be likened to a submerged body in a storm at sea. The storm is produced by the regions of alternating compression and tension which continuously sweep through the region occupied by the organism. As in an ordinary storm at sea, the waves roll by, but from the point of view of the water itself, each droplet is moving to and fro at any instant about some position of equilibrium. Precisely because we are dealing with wave motion, a neighboring droplet will also be engaged in its little reciprocating dance at the same time, but will be reaching its extreme positions at times different from those of the other droplet. The only way in which this progressive phase difference of the various reciprocal motions may co-exist is by suitably stretching and compressing various regions at any instant so as to maintain the passing wave. If the phase difference from one droplet to the next is very small, the stretching and compressing effect is also small compared with the case where the phase difference is much greater. But the rate at which the spatial phase of the wave alters from point to point depends on the wave length. In fact, for a plane wave of ultrasound in a liquid, the local acoustic stresses (i.e., local excess pressure or tension), P, follows the simple formula:

Abstracted with the permission of Ultrasonic Systems, Inc.

$$P = P_o \sin \frac{2\pi}{\lambda} (X - Ct) \tag{1}$$

where P_o is the peak excess compression or tension produced at a point in the medium, λ is the wave length and C is the speed of the wave ($C = f\lambda$). X represents an arbitrary plane in the liquid perpendicular to the plane wave. In order to give mathematical expression to our qualitative amoeba-eye view above, imagine a photograph taken at a time, $t = o$, in the medium. Then

$$P = P_o \sin \frac{2\pi X}{\lambda} \tag{2}$$

and the angle $2\pi X/\lambda$ gives the space phase of the wave all through the liquid at time $t = o$. For example, if one end of our amoeba happened to be at $x = o$ at this time, and if the other end of the amoeba were at $x = l$ (l = instantaneous length of the amoeba), then the amoeba would experience no excess pressure at $x = o$ (since by equation (2), $P = o$). But at $x = l$, its other end, there would be a pressure differential of

$$\Delta P = P_o \sin \frac{2\pi l}{\lambda} \tag{3}$$

Obviously, the magnitude of ΔP depends on the relative magnitude of l with respect to λ. It will only be indicated here that ΔP will be a maximum when $l = \lambda/4$, and for still smaller values of l, P will diminish. For larger values of l, ΔP will pass through one or more maximum values within the organism. In any case, it is evident that the frequency of the ultrasonic energy plays an important role in the possible effects that may be produced. We will hereafter indicate the time-like effect of frequency by computing the high acceleration fields associated with relatively weak ultrasonic vibrations.

There is no question that we are dealing here with a complex of interrelated phenomena, because we know from the literature that ultrasonic fields produce physical, chemical, electromagnetic and biological effects. The extended elementary presentation in this chapter is made chiefly because there is the belief that the low-intensity ultrasonic spectrum, especially in the kilohertz region, is a rich area for exploration by workers in the biological area generally.

Ultrasonic Motor

The emphasis in this discussion has been on the nature of the ultrasonic energy produced by ultrasonic motors, as a background to help workers to develop a better feel for this type of energy as a tool in research. The material has been presented at an elementary level, and no effort has been made to present the more complex phenomena arising from strong vibrations in producing non-linear effects other than cavitation. Material on ultrasonic pumping, microstreaming, and radiation pressure may be readily found in the literature. What the review has strived to bring forward in some detail is the fact that a more definitive study of phenomena, particularly biological, in the presence of ultrasound requires the availability of a suitable energy source, which boils down to having at hand means for obtaining vibrations of various ranges of both *frequency* and *stroke*.

To permit the researcher to conduct his own ultrasonic experiments, it is important that he know more about the ultrasonic motor itself. In all instances the motor comprises a transducer, a transmission system and an output tool. The transducer is a device for taking electromagnetic energy, supplied by a high-frequency alternating current, and transforming it into mechanical vibrations of high frequency. Early research was performed with motors embodying piezoelectric or magnetostrictive transducers. These early motors did not use sophisticated transmission systems, and consequently the "zone of motion" (ZM) available was quite limited.

A typical system includes a motor at 10 kHz, 20 kHz, and 30 kHz, of the piezoelectric type. Each motor has a housing within which is assembled a sandwich of piezoelectric ceramic plates (usually of the lead zirconate type) bolted between metal end pieces. This composite structure protects the otherwise fragile ceramic plates in ordinary handling, and at the same time provides a compressive bias that allows vigorous vibrations to be set up without subjecting the plates to tensile stresses (to which the ceramic plates show little active resistance). To this bolted sandwich transducer a transmission member may be rigidly attached. The transmission member is designed to provide amplification of the output stroke of the transducer and is also proportioned so as to have the same fundamental resonant frequency of vibration as the transducer. The whole structure after assembly will yield maximum response for a given power input when the input high-frequency current matches the mechanical resonant frequency of the transducer transmission line structure (this includes whatever horn is needed at the output end of this motor).

Because significant output of these ultrasonic motors will be available only at a design resonance frequency, it follows that currently available motors are single-frequency sources of ultrasonic energy. It is also important to remember that an electromagnetic converter is needed to operate the motor. The cost of the converter is substantially greater than the cost of the motor itself. Until recently, if a number of frequencies were desired to ascertain the frequency effect in a given research problem, one would normally require a motor-converter combination for each frequency.

A brief review of the properties of the output of an ultrasonic motor is appropriate, since it is just this output that is utilized by the research team, and so the characteristics of the output are more important than the technical intra-industry details by means of which the ultrasonic motor-converter system is put together and designed for the desired output.

We start by concentrating our attention on the reciprocating output surface of our ultrasonic vibrator or motor in order to review the simple characteristics of such motion. First of all, this reciprocating surface has a definite area, and this area in one complete reciprocation, or period, sweeps out a definite volume whose value is just the surface area times the total stroke of the reciprocation. In sweeping out this volume, the output surface of the motor passes through a peak velocity and then passes through an extreme point of instantaneous zero speed, but with peak acceleration. Summing up this description of the output motion of an ultrasonic motor, we may say that there are six interrelated quantities that are useful in assessing the motor's behavior. These are:

Area of output motor surface: S

Total linear stroke from one extremity to the other of one reciprocation: s

Peak velocity of motor surface: v_{max}

Peak acceleration of motor surface: a_{max}

Time or period of one complete reciprocation: T_o

Frequency of vibration of the motor surface: f_o

Elementary physics requires the following interrelationships:

$$v_{max} = \pi f_o s \qquad (4)$$

$$a_{max} = 2\pi f_o v_{max}$$
$$= 2\pi f_o^2 s \qquad (5)$$

$$T_o = \frac{1}{f_o} \qquad (6)$$

$$V = Ss \qquad (7)$$

where V is the volume swept out during one stroke, s, and we call this volume the "zone of motion" of the motor output.

These simple relations are the keys to understanding most of the effects, many of which are unique, which an ultrasonic motor can produce. There is a fifth relationship, which expresses Newton's Second Law of Motion and may be written:

$$F = ma \qquad (8)$$

where F represents the net external force on mass, m, and a is the acceleration produced in m by F. Of these five relations, equation (5) is probably the most important for giving insight into the biological effects that may be produced by ultrasonic motors. Equation (5) asserts the way in which the frequency, the peak acceleration, peak speed, and peak stroke of an ultrasonic motor output are connected.

An ultrasonic motor is distinguished by the fact that its frequency of reciprocation is generally above the limit of human hearing, or above about 16,000 vibrations per second, while its peak stroke is generally microscopically small, usually expressed in mils (thousandths of an inch). In order to see what kind of ballpark we're playing in as to the magnitude of these quantities, suppose we take a commonly found case for ultrasonic motors, namely, a frequency of 20,000 cycles per second and a peak stroke of 2 mils. In this case, we can calculate the peak speed, v_{max}, and the peak acceleration, a_{max}, from equations (4) and (5). Doing so gives:

$$v_{max} = 10.5 \text{ ft./sec.}, a_{max} = 41{,}000g \tag{9}$$

where

g = acceleration of gravity = 32.2 ft./sec.^{2}

f_o = 20,000 cycles/sec.

s = 2 mils

Equation (9) tells us that the output surface, S, of our ultrasonic motor reaches a peak speed of 10.5 ft./sec. or about 7 mph, while it also reaches a peak acceleration of 41,000 times the acceleration of gravity. In other words, under the prescribed conditions of frequency and stroke, the ultrasonic motor describes an invisible zone of motion, never attaining more than horse-and-buggy speed, but with a peak acceleration which is enormous compared with gravity. This unique state of affairs cannot be duplicated by any other known means, and herein lies the uniqueness offered to the biological research team in probing their area of study. What does it mean in terms of new possibilities for biological research? To pursue this question we must return to the zone of motion being swept out by the ultrasonic motor output area, S.

Figure 3.1 shows what we have been discussing. P_1 and P_3 are extremities of the motor stroke, and P_2 is the mid-point of peak speed, v_{max}. Just to consider a specific situation, suppose the output area, S, is submerged in a liquid and the motor is started. Due to successive contraction and expansion of the motor output section, the area, S, will oscillate around the point P_2, between the extreme points P_1 and P_3. With the motor output submerged in the liquid, the question arises as to how the liquid will respond to this zone-of-motion (ZM) reciprocation.

For example, if S is at P_3, then in the next instant it will move upward with an acceleration of 41,000g. For the liquid to be able to follow S, it must be able to attain this same high acceleration at P_3. But the strength of most (if not all) liquids is such that it cannot sustain such an acceleration in producing a tension at the upward-moving surface. Therefore, the liquid surface at P_3 breaks into a number of small fissures or micro-cavities. These cavities are predominantly small bubbles containing essentially nothing but some vapor molecules of the parent liquid. Furthermore, in the next 25 micro-seconds (remember, the period of reciprocation equals 1/20,000 second or 50 micro-seconds), the surface,

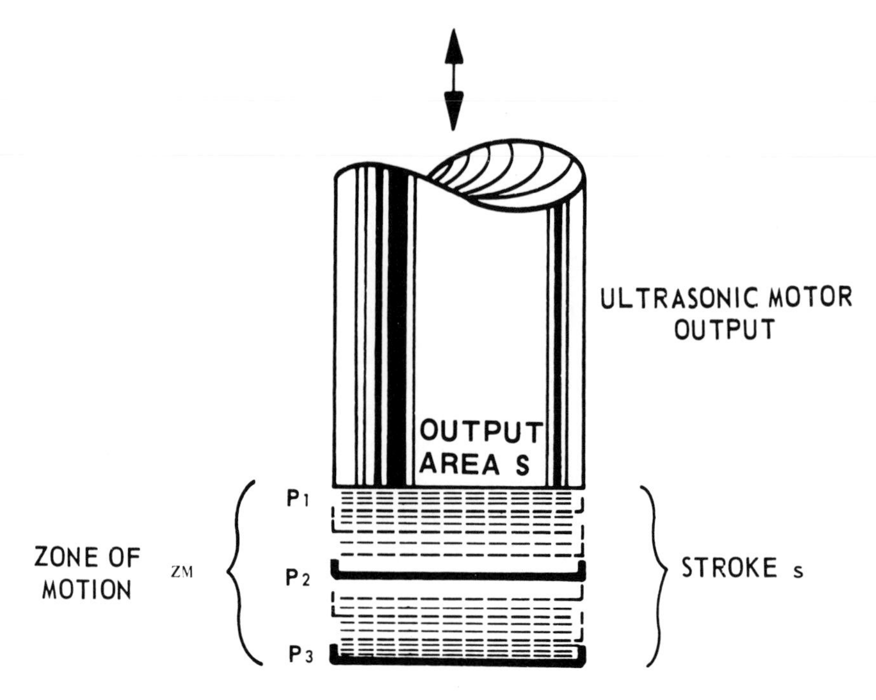

Figure 3.1. Rates of acceleration up to 41,000 times the acceleration of gravity are obtained by ultrasonic motor output surfaces in reciprocating through their zones of motion (ZM).

S, starts its return sweep and forces the rapid collapse of the cavities produced. This phenomenon proliferates 20,000 times each second, in consequence of which literally millions of miniature shock waves (like micro thunderclaps) occur every second. Thus the liquid, being processed, is filled with what we call "cavitation."

The output of the vibrating motor is always an area, S, which is reciprocating at some frequency, f_o, with a total excursion or stroke of magnitude, s. Thus Ss is a volume, V, swept out per stroke of the motor by the output area, S. We will call this the "zone of motion" of the motor. This zone of motion, V, is repeated $2f_o$ times each second, since there are two complete strokes for one complete reciprocation of the motor vibration. Thus, if the motor output is operating in a liquid, the total liquid volume displacement per second is as follows (V_D = liquid volume displace per second):

$$V_D = f_o V \qquad (10)$$

$$= f_o Ss$$

Note that we do not say $2V_D$, because the liquid is displaced only on the forward stroke of the reciprocating motor output section. If, however, we wished to know the total distance traveled in one second by the output section, we would take $2s$ for each complete oscillation, and then we would get, by adding this distance for f_o vibrations, the distance, d,

$$d = 2f_o s \qquad (11)$$

Of course, this distance is to and fro. Thus, on the average, there is no net motion of the output section itself, but the effects on the liquid are additive and alternating in time, and so the quantities in equations (10) and (11) (V_D and d) do give a measure of performance for an ultrasonic motor. In order to appreciate more the possibilities inherent in these quantities, we need to look a little further into the character of these to-and-fro motions of the motor in the liquid.

First of all, the motion is the type which, in mechanics, is called "simple harmonic." This kind of motion is elementary and well known in many vibration phenomena, and takes place according to a simple sinusoidal relation between the linear displacement of the motor and the time, t. For example, the linear displacement, x, at any time, t, of the motor output section may be represented as:

$$x = \frac{s}{2} \sin 2\pi f_o t \qquad (12)$$

where s is the stroke and f_o the frequency of the motor. From this equation, it is easy to derive the information as to the peak speed, v_{max}, and the peak acceleration, a_{max}, reached by the motor during each stroke, s. The results are:

$$v_{max} = \pi s f_o \qquad (13)$$

$$a_{max} = 2\pi f_o v_{max} \qquad (14)$$

$$= 2\pi^2 f_o^2 s$$

Ultrasonic Motor Limitations

It is clear that all the basic mechanical properties of the ultrasonic motor output are determined once the stroke, s, and the frequency, f_o, are fixed. The frequency of operation is a matter of motor design, and the stroke depends on motor design and power transfer to the motor from the generator. As the power to the motor goes toward increasing the output stroke, the peak stresses in the motor structure go up. No matter what motor design is used, there will always be some peak stress (and therefore strain) which is *not* safe to reach.

It is a law of the ultrasonic motor that the peak stress it can sustain is proportional to the peak velocity it can deliver at its output. This law is *independent* of the *frequency* of *operation*. Thus, for a given type of design, one can assume some upper level of peak velocity output, v_{max}, which can be attained no matter what the frequency. Turning to equation (4), we have:

$$v_{max} = \pi f_o s \tag{4}$$

Therefore, if v_{max} is a constant, $f_o s$ is a constant; and it follows that if one wishes to operate at a *higher* value of f_o, one must accept a *lower* value of reciprocal stroke, s. For example, if a certain motor type can put out, say, 5 mils at 20,000 cycles/sec., then at 10,000 cycles/sec. it can put out 10 mils, and at 100,000 cycles/sec., it can only put out 1 mil. Knowledge of this fact is especially important when one is endeavoring to discover whether some known biological or other effect depends principally on frequency or stroke. If high stroke is needed, low frequency is indicated; while if high frequency is needed, the stroke is of lesser importance. Finally, if the effects sought depend primarily on output speed, v_{max}, then frequency is unimportant, since v_{max} depends only on the peak stress in the motor and not on frequency.

There still remains the basic quantity, a_{max}, the peak acceleration. According to equation (5):

$$a_{max} = 2\pi f_o v_{max} \tag{5}$$

This relation tells us that the peak acceleration depends linearly on frequency, since v_{max} may be regarded as a constant when you are considering the limits of motor operation. If higher accelerations are important, one should consider using a motor of higher frequency. It must be emphasized that the considerations of this section hold true only when we are considering the top possible performance of ultrasonic motors.

It is hoped that this section makes it clear that in order to obtain a more complete picture of the possibilities of ultrasonic effects in biomedical and other work, it is necessary to have a range of operating frequencies available. The projected system provides a multifrequency motor converter combination, wherein one converter serves to operate three motors in three different frequency ranges.

Let us now return to the biological and biochemical worlds and consider the use of ultrasonic energy therein. In the first place, for organisms in liquids (which covers practically all living systems), the use of ultrasonic energy divides into two broad categories. One that is currently in widespread use corresponds to a motor output which produces a cavitation field in the liquid. For any liquid biological system there is a threshold ultrasonic motor stroke at a given frequency, such that for strokes greater than the threshold, cavitation will occur. Below this threshold, no cavitation occurs, but sinusoidally alternating compression and tension proceed in the liquid.

The equipment illustrated in Figure 3.2 provides a portable source of

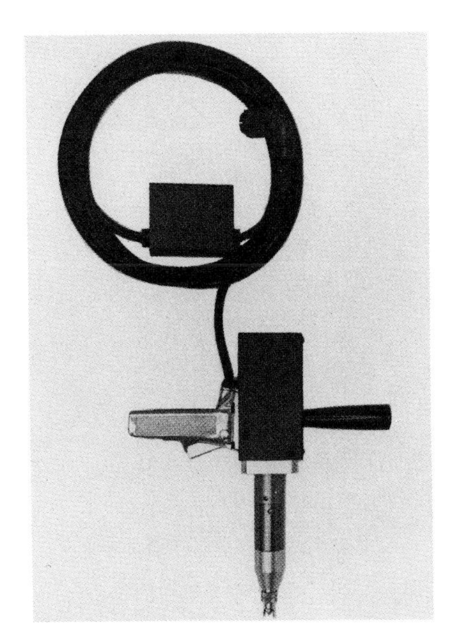

Figure 3.2. Portable ultrasonic welder.

focused ultrasonic energy with three primary variables: *frequency, intensity* and *time*. This provides an ultrasonic laboratory which is available to the scientist, researcher and lab technician to permit him to be economically equipped to perform experimental work on various applications and to conduct (1) fundamental research; (2) pilot plant feasibility study; (3) application and product development. The basic area of application of the equipment is divided between biological, chemical and physical uses of ultrasound.

Chapter 4

Equipment and Methodology

There are two stages in developing a basic understanding of ultrasonics and its application to practical problems in the field of packaging. The first stage is developing a comprehension of the actual components or hardware that is used. This might be referred to as the principles of ultrasonics. The second stage is the knowledge of the technique of using this equipment to solve specific problems. This might be referred to as the methodology of ultrasonics. This chapter, therefore, presents a basic outline of the equipment and methodology of ultrasonics.

Ultrasonics literally means "beyond sound," i.e. sound above the audible range. Taking 18,000 Hz (cycles per second) as an approximate upper limit of human hearing, ultrasonics refers to sound above 18,000 Hz.

To eliminate an irritating noise problem, 20,000 Hz was selected as the operational frequency used by most manufacturers. Equipment has been constructed that operates in the 10,000 Hz region; however, acoustic isolation is required either in the form of sound deadening chambers or headsets.

Principles

Most conventional plastics assembly systems comprise four basic elements: power supply, converter, horn and stand.

Power Supply

The power supply or generator transforms 115-volt, 60-Hz electric energy into high-frequency electrical energy at 20,000 Hz. Power supplies are rated in watts of output, and typical units range from 100 to 2000 watts.

Converter

The converter receives electrical energy from the power supply and converts this electrical energy into mechanical vibratory energy. There are two types of converters presently used in plastics welding: piezoelectric and magnetostrictive.

Piezoelectric. Piezoelectric converters employ polycrystalline substances, such as barium titanate or lead zirconate titanate, which when polarized exhibit electrostrictive behavior (change dimension when electrically excited). When the piezoelectric material is excited by the 20,000 Hz electric current from the power supply, it mechanically vibrates at 20,000 Hz.

Magnetostrictive. Magnetostrictive converters employ ferromagnetic materials such as nickel, vanadium, Permendur, and ferrite, which change dimensions when magnetized. When a 20,000 Hz alternating current flows through the excitation coil, an alternating magnetic field is developed, and the magnetostrictive core mechanically vibrates at 20,000 Hz.

Horns

To perform work it is necessary to transfer the mechanical vibration to the workpiece. The primary function of the horn is to deliver the energy to the desired location. The vibratory power must also be delivered to the workpiece at the proper force-amplitude ratio.

A horn is a half-wave-length section designed to resonate at 20,000 Hz (20,000 cycles per second). The shape of the horn determines the gain or magnification, i.e. the ratio of output amplitude to input amplitude. Its cross-section is modified to provide an output shape as desired, producing either high amplitude-low force energy or high force-low amplitude energy. Motion, or vibration, is produced by expansion and compression in the horn material. The mechanical vibration produced by the converter is axial in direction, and the total excursion of one complete cycle is in the order of 0.0008 in. Vibrating horns are therefore stressed. For any given horn shape, the stress is proportional to the amplitude of vibration; therefore, every horn has an amplitude limit. When the diagonal of a horn exceeds 3½ in., radial or cross-coupled stresses are produced that require special design considerations. Shapes not balanced with respect to the longitudinal axis of the horn tend to produce flexure modes of vibration. Sharp notches produce stress concentrations that have decreased allowable amplitude.

It is possible to *stall* a horn. When this occurs the horn attempts to draw more power than the power supply can produce. The ideal force-amplitude ratio varies for each application; however, with most welding applications, the more power that can be delivered to the joint interface, the faster the assembly procedure is accomplished. The *stall* condition does not occur just past the peak-power or maximum-loading point, but rather after a gradual decrease in power has occurred. The maximal power output (a) can be determined from the meter located on the power supply or stand. This is an indication of percentage of power drawn. When a maximal reading of 100 has been obtained, increasing the amplitude and/or pressure will diminish the power output (B) until a total stall condition exists (C).

Ideally, horns are made of materials that are strong and have good acoustical characteristics. Titanium alloy has the best acoustical properties of all high-strength alloys. Aluminum alloy has excellent acoustical properties, but because of its lower strength and hardness, it is subject to wear and fracture of highly stressed designs. Monel metal is a high-strength alloy with acoustical properties less desirable than titanium. The advantages of Monel is that hard-surface materials may be brazed to the output end of the horn, as opposed to titanium alloy which is not suitable for brazing. The poor acoustical properties of steels limit their use to low-amplitude horns for inserting. Tool steels may be heat-treated to produce very good abrasion resistance.

Ultrasonic horns are available in a variety of shapes, each designed for specific applications and categorized by its name. Horns have been manufactured to conform to parts of every imaginable shape and generalization becomes most difficult. It must be remembered that the designs which follow are most elemental and provide only a basis of horn design.

Stepped Horns. Stepped horns consist of two quartered-wave-length sections of different but constant cross-sections, joined together at the node with a relatively small radius. This type of horn produces very high gain, but the horn material becomes very highly stressed when driven to high amplitudes. The end of the horn may be tapped to accept replaceable tips.

Exponential Horns. Exponential horns have very desirable stress distributions, but very low gain factors. Applications using exponential horns are generally limited to driving metal inserts into plastics. The end of the horn may be tapped to accept replaceable tips.

Catenoidal Horns. Catenoidal horns are a practical compromise between a step horn and an exponential horn. This type of horn is most suitable for plastic welding and staking of small parts. The end of the horn may be tapped to accept replaceable tips.

Rectangular Horns. Rectangular, or bar, horns have many configurations. Small rectangular horns are usually stepped. Larger bar horns are sometimes tapered at the larger end. The maximum width for a non-slotted bar horn is 3½ in. Slotted bar horns may be made up to 12-in. wide as required for the particular welding application.

Circular Horns. Circular horns can be made hollow or solid. Non-slotted circular horns are made up to 3½-in. in diameter. Slotted circular horns may be made as large as 10-in. in diameter for special applications.

Amplitude Transformers. An amplitude transformer is a type of horn that is used only to alter the amplitude. It is connected between the converter and horn and makes it possible to achieve variations in amplitude that could not be achieved with the horn alone. The combination of pressure, amplitude, the part and its composition, and the duration of ultrasonic exposure time will determine the amount and degree of heat produced. It is important to realize that if amplitude (horn excursion during one cycle) is increased, there is a sacrifice of force capability. For purposes of illustration, a high-amplitude horn could be compared to driving a car in third gear, where high speed is required with very little available torque. Conversely, a low-amplitude horn has great torque capability, and can be compared to a car in first gear. It is possible to *stall* a high-amplitude horn by operating it under high-pressure conditions, just as it would be easy to stall a car motor if started up a steep hill in third gear.

Typically, six standard amplitude transformers are available, offering a reverse ratio of 1:0.04 to a positive ratio of 1:2.5. Amplitude transformers that increase amplitude are called *boosters*.

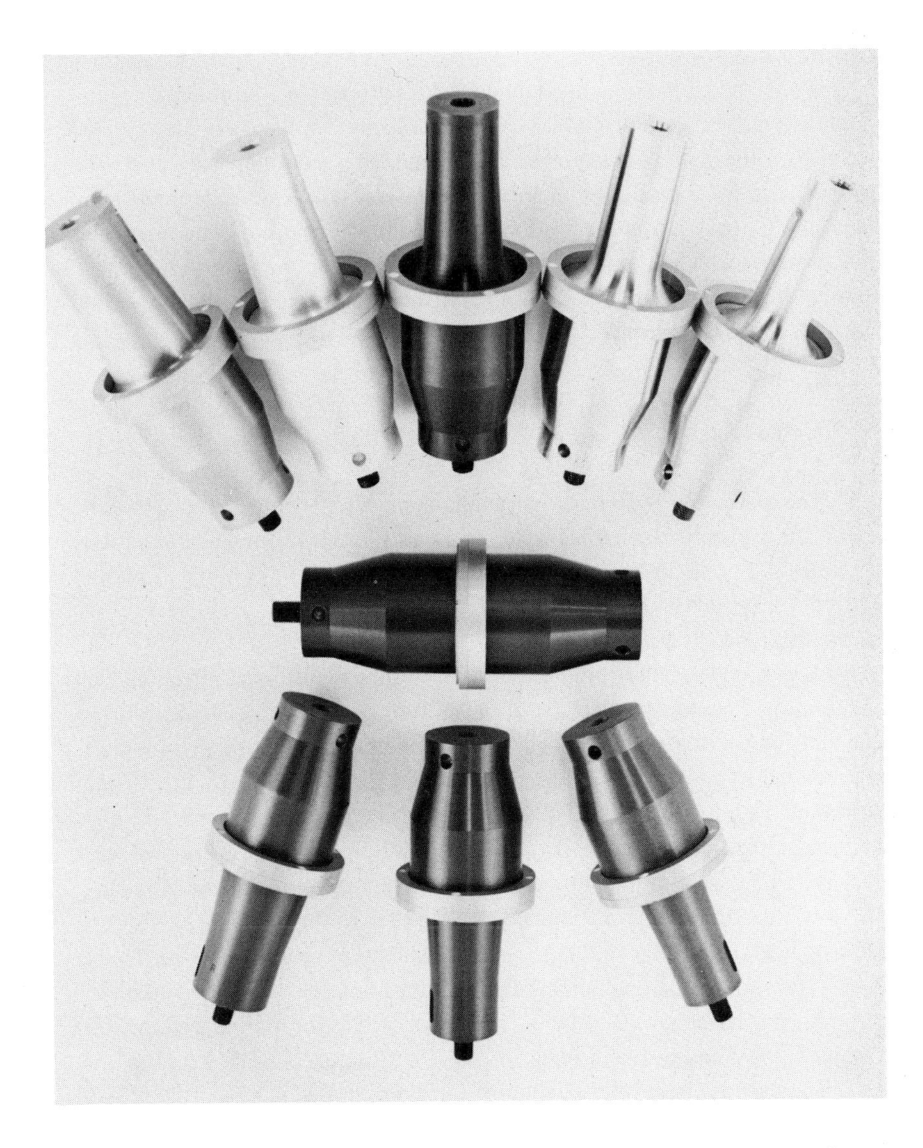

Figure 4.1. "F" amplitude transformers, varying in ratio from approximately 1:0.4–1:3.5.

Stand

The stand, or welding press, houses the converter and horn in a rigid mounting which moves up and down pneumatically, applying a predetermined pressure on the part. Controls such as ultrasonic exposure time and dwell time (pressure applied after discontinuance of ultrasonics) may be located either on the stand or at the power supply. One manufacturer has produced a new generation of stands which incorporates the entire power supply within the stand housing. Such construction permits all controls to be placed on the face of the unit that actually performs the assembly function.

Methodology

The method of applying specific ultrasonic equipment to practical problems is the bridge between ultrasonic theory and its wide commercial usefulness. There are five basic phases in the operations involved in the bonding of plastics and other materials with ultrasonics: (1) vibration; (2) friction; (3) heating; (4) melting; and (5) welding.

Acoustic properties vary with different plastics. Just as a bronze bell will ring because it is acoustically a good transmitter of vibrations as compared to a bell made of lead, so is general-purpose styrene a good transmitter of vibrations as compared to low-density polyethylene. When the vibrations from the frontal surface of the horn are applied to the polyethylene part, for example, the energy will be absorbed within the plastic, and as a result, insufficient energy will be transmitted to the joint area to melt the plastic. Conversely, styrene transmits the vibrations from the point of ultrasonic contact to the joint quite efficiently; sufficient motion is created at the joint interface to momentarily bring the temperature to the melting point. Charts in this book list the various plastics in the order in which they efficiently transmit vibrations for remote welding. (See Tables 7-2, 8-3 and 8-4; and Figure 12.9).

The distance between the point of ultrasonic contact and joint area will also affect the weldability of a given part. Since even the best plastics are poor transmitters of vibrations compared to metal, more energy will be available if the vibrations travel only an inch from the point of ultrasonic contact to the joint than if they travel 12 inches. Welds have been made on styrene thermo-walled tumblers where the vibrations are applied to the base and must travel the entire

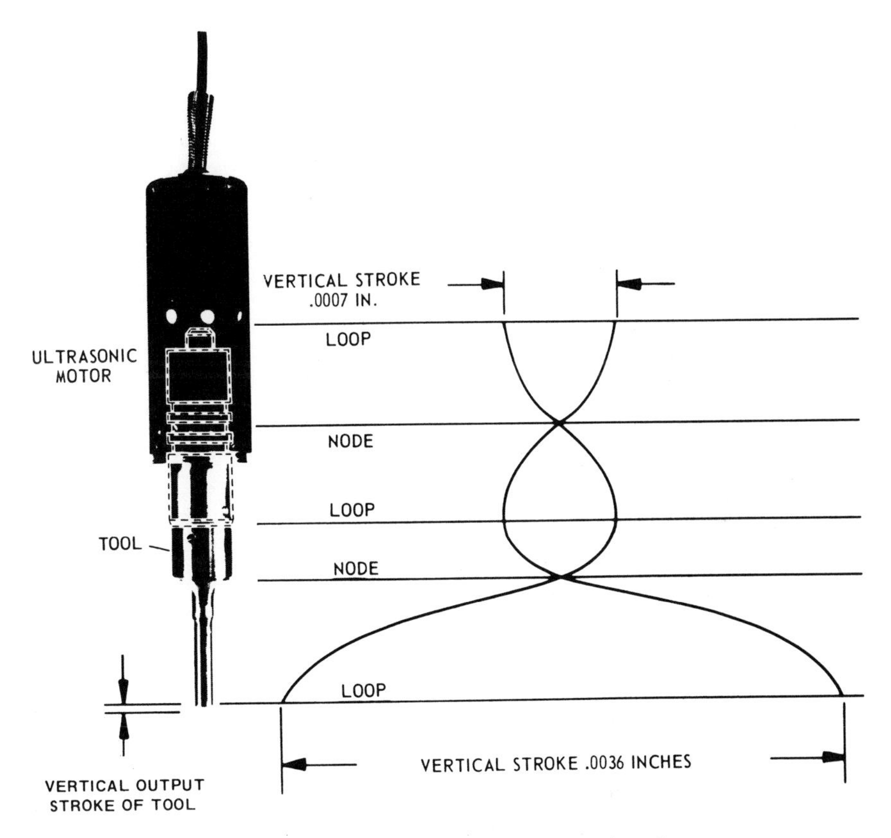

Figure 4.2. Ultrasonic motor travel cycle.

8-in. length to the joint area near the mouth of the tumbler ("far-field" welding). On such an application, a welding time of 2.0 seconds may be required, whereas if the vibrations can be applied at the mouth of the container so there is only an inch or less distance to the joint line, the exposure time could be as short as 0.5 seconds ("near-field" welding).

Applied Methodology

With very few limitations, almost any requirement of ultrasonics can be provided with designed methodology. The concept of application must of course have the prerequisites of materials, sizes, and capabilities to incorporate the

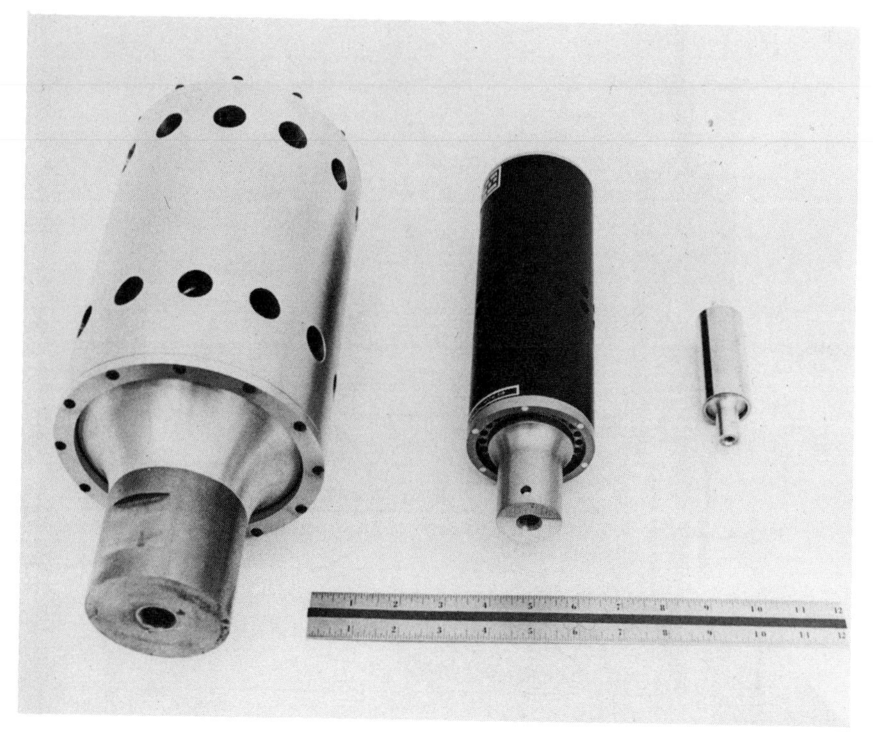

Figure 4.3. Size differential between a 10 kHz (largest), a 20 kHz, and a 40 kHz converter.

proper horn design. The methodology may then be designed to handle applications from single to multiple requirements.

Welding

Joint. The key to successful joint design is to keep the initial contact area as small as possible. For example, a simple butt joint can be optimized for ultrasonic welding with the addition of a small protrusion called an energy director. Since there is less plastic contact on a square-inch basis, melting will occur much more rapidly, allowing that the height of the energy director is one-tenth of the wall thickness, whereas its width is one-fifth of the wall thickness. The same principle is used for the step joint and a tongue-and-groove joint. The molten material from the energy director is calculated to fill the interface area. Since the welding cycle occurs within a fraction of a second, no degradation of the plastic or crazing occurs.

Practical considerations suggest a minimum height of 0.005 in. for the energy director. Where a height greater than 0.020 in. is indicated, two or more directors should be provided with the sum of the heights equalling the suggested dimension.

Shape. The general shape of the parts to be assembled can readily affect welding. Symmetrical parts may weld better and faster because they transmit the acoustics very uniformly to the joint area when far-field welded. An example would be a dodecahedron, where the joint area is not parallel to the tip of the horn, but because of geometric shape, the joint welds readily.

Odd shapes, such as many toys, automobile parts, and electrical devices, can also be welded when they represent good structural configuration and permit ideal placement of the horn for transfer of the ultrasonic energy.

Near- or Far-Field. Any welding where the location of the joint is less than 1/4 in. from the horn is considered near-field. Far-field welding takes place more than 1/4 in. from the horn.

Energy-absorbing plastics such as polyethylene or polypropylene can usually be welded only in near-field. It can be easily seen what occurs in far-field welding of these materials: the energy is absorbed in the flexing action of the substance and not delivered to the joint.

Far-field welding can be accomplished with success on weldable plastics. The general requirements are structural integrity of the parts, good joint design, rigid material composition, and horn placement in an area where the energy can be directed to the joint.

Nesting. Holding the parts in alignment during welding is another aspect that must be taken into consideration. Nests to hold the parts in proper alignment can be as simple as using two pins in a steel block to locate the part under the horn. More elaborate systems involve clamping and alignment mechanisms to hold tolerances within a few thousandths of an inch. Pressure-tight joints, for example, must be perfectly shimmed so that the horn exerts equal pressure around the entire periphery of the part to give uniform welding.

It is important that the nests be constructed so that there is no lateral movement of the part under the horn at the moment of welding. This prevents the horn from "walking," which can mark the surface of the part. Mirror-like surfaces can be contacted without leaving horn marks if proper precautions are taken.

Whenever contoured surfaces need to be contacted, the face of the horn must be shaped to conform to the part to prevent marking.

Inserting

It is possible to melt plastic and flow it into knurls, flutes, and undercuts to provide a mechanical bond rather than a chemical bond. A metal insert can be forced into a piece of thermoplastic material and mechanically held in place by the plastic that has been displaced into the undercut area. As the result of the insert being vibrated against the plastic, heat is generated and momentarily melts the plastic immediately adjacent to the insert. Using this technique, several parts can be inserted simultaneously into plastic within a fairly small area.

Although it is not absolutely necessary to pre-mold the hole, it is usually found advantageous to do so. Since the volume of molten plastic displaced by the insert must have some place to go to prevent unsightly flash, it is necessary to dimension the hole properly. The insert will also follow the path of least resistance. The molded hole therefore acts as a pilot for the insert, which is not the case when the insert is driven into solid plastic.

Design Advantages of Ultrasonic Insertion. (1) *High Strength:* uniform melting and resolidification provides a mechanical grip which equals that of molded-in inserts; (2) *Minimum Stress:* minimum outward stresses are created when the insert is loaded in service, and smaller bosses are required; (3) *Tolerates Variation in Hole Size:* equal performance is achieved with variations in hole diameters of four to five times normal molding tolerances.

Staking

Another important aspect of ultrasonic assembly is staking. Since there is no joint in a part to be ultrasonically staked, the friction and resultant release of energy occurs between the surface of the horn and the plastic.

Poor coupling is achieved by keeping the contact area between the horn and the plastic as small as possible. The horn is usually contoured to meet the specific requirements of the application.

By shaping the face of the horn, the flow can be controlled to almost any configuration that is required. Knurled patterned, dimpled, or rounded-head horns can be incorporated.

Two big advantages of ultrasonic staking over conventional convection heating techniques are evident. First, the speed at which the head is formed is normally much faster than using a hot-iron approach, since the mechanical energy travels into the plastic more readily, compared to heat-energy transfer rates. Many single-head applications require as little as 0.2 seconds. With plastics having high melting temperatures, the time saved using ultrasonic staking techniques versus heat staking is substantial.

The other big advantage of ultrasonic staking is that there appears to be no degradation of the material. This is because the material is made sufficiently soft to flow just before it reaches its melting temperature. The head of an ultrasonically staked part will have as much strength as the parent material itself.

It has been found that the thermoplastic material will not adhere to the face of the horn, and since the horn is relatively cool as compared to the molten stud, as soon as the ultrasonic exposure is terminated, the plastic solidifies to form a tightly locking head.

Sonic Sewing

Synthetic fabrics such as nylon, polyester, polypropylene, modified acrylics, some vinyls, urethanes, and many synthetic blends of up to 35% natural-fiber content can be bonded (see Chapter 15). Materials are fed between the vibrating horn and stitching wheel; the fibers are compressed into intimate contact as determined by the pattern of the stitching wheel, and the resultant frictional heat melts the fibers into a rigid solid seam. The obvious advantage is that the heat for melting is created within the material itself at a rate not possible with other methods dependent upon thermal-conductivity to effect a bond.

Advantages of Sonic Sewing. (1) No needle, thread, or bobbin; (2) no breakage or adjustment of (1), above; (3) no inventory or color-matching of thread; (4) no thread to trim after seam is completed; (5) variety of stitch patterns quickly set up; (6) reduced operator training; (7) fewer adjustments; (8) less hazard (no needle punctures); (9) unlimited throat clearance available; (10) single and multiple stitching available in same machine.

Disadvantages of Sonic Sewing: (1) Not applicable to natural fibers (cotton, wood, linen, paper, etc.); (2) not applicable to rayon and other cellulosics.

Section II

Equipment and Processing

Chapter 5

High-Intensity Systems

The high-intensity source of ultrasonic energy consists of two major elements: converter and motor.

Converter. A solid-state electrical converter that takes 60 Hz power from the electrical outlet and converts it to 20 kHz (20,000 cycles per second) to drive the motor.

Motor. A motor is comprised of a transducer, horn transmission line and tip element.

Transducer: Receives electrical energy from the converter and transforms it into mechanical vibratory energy.

Horn: Essentially a transmission line to direct the mechanical vibratory energy from the transducer into the most useful form at the tip.

Tip: Performs extensionally with kinetic oscillations sufficient to deliver the desired effect directly into the material being treated. This accessory may be interchanged with an assortment of more sophisticated tip elements of varying configurations.

Uses of High-Intensity Ultrasonics

High-intensity ultrasonics may be used in any process requiring extreme intensities of ultrasonic energy. Typical uses in the biological, chemical and medical fields include:

Homogenization
Emulsification
Automization
Oxidation

- Extractions
- Dispersion
- Deaerating
- Degassing
- Defoaming
- Disintegration
- Solubilization
- Catalysis
- Cell Disruption
- Cleaning
- Sterilization
- Chemical Activation
- Mixing

Function of High-Intensity Systems

Figure 5.1 illustrates how the tip produces a zone of motion (ZM) during oscillation. This is the relatively simple yet all-important phenomenon that produces the great variety of uses for which high-intensity systems are employed.

The ZM is microscopic; its stroke ranges from a few microns to several thousandths of an inch. However, though the motion is minute, the total strokes per second (2 times the frequency of the motor) enable the tip to displace 1.2

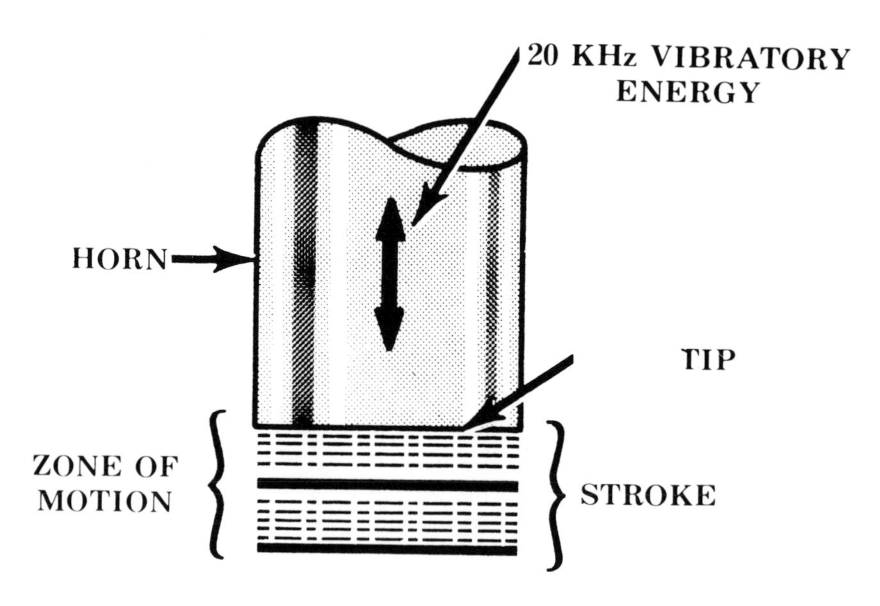

Figure 5.1. The zone of motion (ZM) is the area under the horn through which it moves when it is activated.

liters of volume per square inch of tip each second. Based on a stroke of 0.0035 in. at 20 kHz, the tip travels a distance of 12 feet in one second.

During oscillation, each stroke of the tip is generating a peak acceleration of over 72,000 times the acceleration of gravity. Herein lies the unique function of the high-intensity unit, the dynamics of which cannot be attained by any other known instrument.

Whether you are disrupting cells, extracting, solubilizing or homogenizing, what your horn does for you is completely determined by the stroke and cross-sectional area of the vibrating tip end.

It must be emphasized that so long as the ZM is maintained at a given level, the effects produced are determined by the action of the tip. It is the magnitude of the ZM that governs the performance of high-intensity ultrasonics in a given application. The ZM rating can thus be defined as stroke times radiating area produced under working conditions, such as liquid immersion.

Zone-of-Motion Economics. As the cost of fuel has a direct bearing on how you heat your home, proper application of the ZM will affect another important consideration: your occupation budget responsibility. The smaller the converter required to produce the desired ZM, the more economical your high-

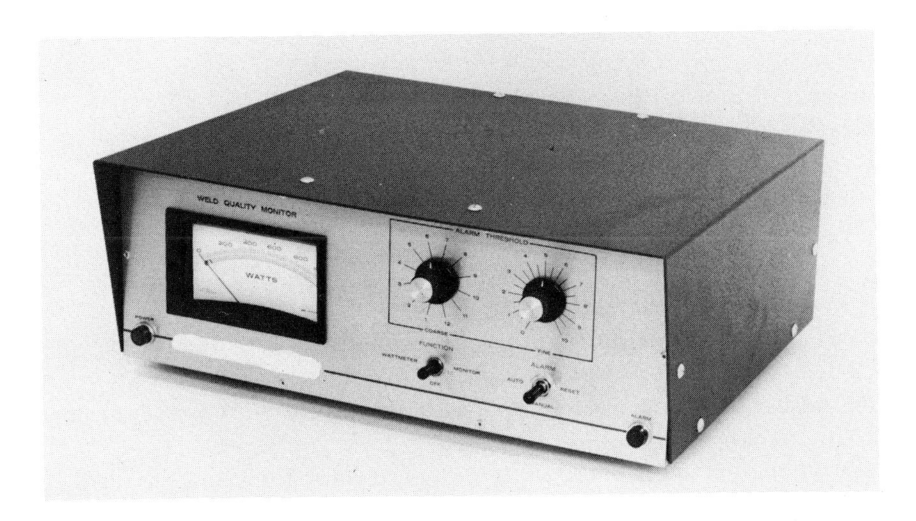

Figure 5.2. A weld quality monitor can be used to measure the power drawn to perform an application, or reject a part that has been unsatisfactorily assembled.

Figure 5.3. Size differential between 8000-watt power supply, 1000-watt power supply, and 40-watt power supply.

intensity ultrasonics will be. Thus a primary reason why high-intensity systems are made in a variety of power levels is to suit varying research needs. It follows, then, if you are considering the use of ultrasonic energy in your work, you will best be served by a system that supplies the ZM you require at minimal power input.

Cavitation. Cavitation is defined as the formation of voids in a flowing liquid as a result of the separation of its parts. In keeping with its role as a multifaceted aid to laboratory research, high-intensity ultrasonics, through its repetitive zone-of-motion generation, is also capable of selectively producing the cavitation phenomenon.

Thousands of vaporous cavities are formed during the contracting segment since the liquid is incapable of following the high acceleration of the tip and therefore breaks into microscopic voids. During the return or compressive tip stroke, the voids collapse in less than 25 microseconds; each void produces

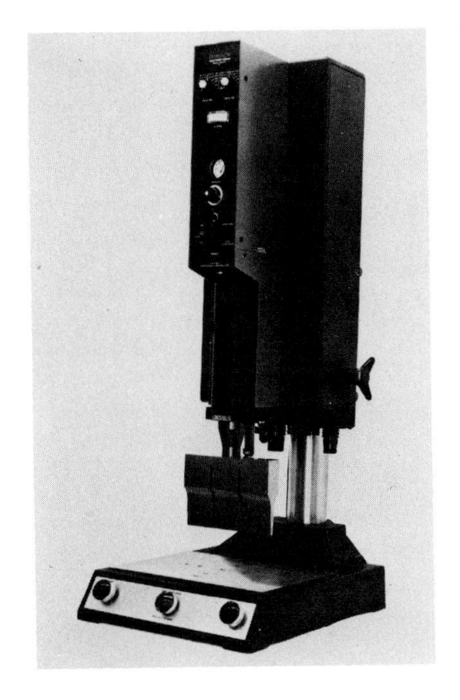

Figure 5.4. The model 4120 welder. It is the highest-powered, integrated plastics welder available, and workspace is conserved since power supply and head assembly are all in one unit.

a miniature thunderclap or shockwave. It is the ZM of the motor which, by the production of cavitation shock waves, creates all of the end uses.

Converters

Combined with the motor, an ultrasonic unit requires a frequency converting generator-oscillator (converter) to power the ultrasonic motor in its frequency range. Converters must be capable of converting either standard electric (60 cycles per second) or battery current into high-frequency energy. By using solid-state components, small electric converters have been developed which deliver current to the motors. By incorporating transistor-type switches and relays into these electronic converters, a minimum amount of moving parts can be designed into converters so that the entire unit, comprising both the motor and converter, will require minimal maintenance.

For peak efficiency, the frequency of the energy delivered by the converter is adjusted to vibrate at the same frequency as the natural resonant frequency of the motor. This adjustment may be accomplished by a manually operated tuning device or by means of special circuitry incorporated into the converter which produces automatic frequency control. There are many presently known ways to make a converter self-regulating with respect to a given motor. This self-regulating feature is desirable because in many applications, when the motor is in use, the use itself may cause small fluctuations in the motor's frequency and thus considerably change the motor's useful output. Self-regulation automatically adjusts the frequency of the converter to the motor.

Motors

An ultrasonic motor is a low-speed, high-reciprocation-rate device, which functions by means of vibratory energy, instead of rotational energy characteristic of ordinary motors. It consists of a transducer, which converts electrical energy into mechanical high-frequency vibration energy, an elastic wave transmission section and a tool or output section for delivering the vibration energy to its destination. In each of these parts, vibrations in the form of elastic waves occur, which cause each small part of the motor to execute minute oscillations invisible to the human eye. Thus, the ultrasonic motor has no gross moving parts, and no bearings or lubricants are needed for such reciprocating devices as are required for ordinary electrical motors. Ultrasonic motors, therefore, require minimal maintenance, making them attractive for many uses.

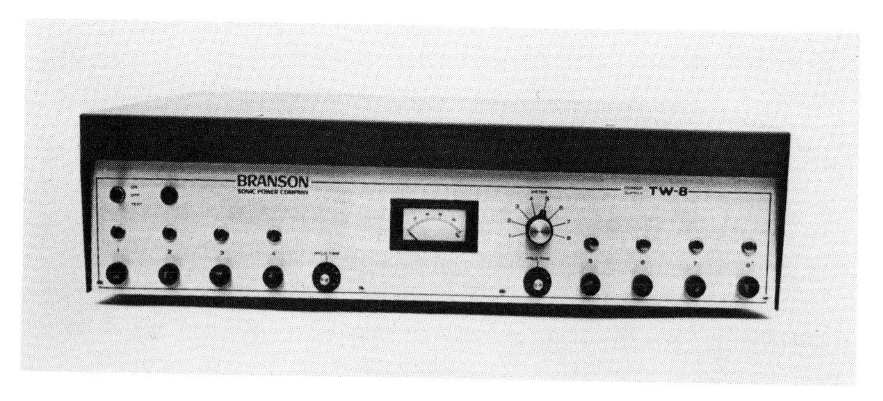

Figure 5.5. The TW-8 power supply can simultaneously power as many as eight welding machines.

Based on $(1\ \text{inch})^2$ cross-section output surface.

Frequency	Stroke (s)	v_{max}	a_{max}	$d = 2v_{max}/\pi$	$V = (Ss)$	$V_D = \frac{(Sd)}{2}$
KHz (1)	mil (2)	FPS (3)	g (4)	cm/sec	cm^3	liter/sec
10	4.0	10.5	20,500	204	.0645	.66
20	2.0	10.5	41,000	204	.0323	.66
30	1.5	10.5	61,500	204	.0246	.66
40	1.0	10.5	82,000	204	.0161	.66
80	0.5	10.5	164,000	204	.0081	.66
160	0.25	10.5	328,000	204	.0040	.66

Units have been selected to assist in visualizing the actual behavior of the motor rather than to adhere to one system.

(1) KHz = Kilocycles per second or Kilohertz.

(2) mil = one thousandth of an inch

(3) FPS = feet per second

(4) g = acceleration units given as multiples of the acceleration of gravity $g = 32.2\ ft/sec^2$.

Figure 5.6. Motor relationships to cross-sectional output.
Reprinted with the permission of Ultrasonic Systems, Inc.

Booster Horns. A horn is a one-half wave resonant metal section designed to transmit mechanical vibrations from the sonic converter to the material being processed. Proper design of the horn is essential for optimal effectiveness on each ultrasonic application. Acoustic properties of the material of construction, length, and distribution of mass all affect resonance and output force-velocity ratios for any given input and load.

Horns are constructed from a special titanium alloy having good acoustical properties while possessing a high strength-to-weight ratio.

The success of welding and staking of plastic or inserting metal into plastic depends upon the proper amplitude of the horn tip. Since it may be impossible to design the correct amplitude into the horn initially because of its shape, booster horns are necessary to either increase or decrease the amplitude to produce the proper degree of melt or flow in the plastic part. The choice of plastic, the shape of the part, and the nature of the work to be performed all determine what the optimum horn amplitude should be.

Six amplitude-modifying booster horns are available—three for increasing amplitude and three for decreasing amplitude. Each horn is anodized with a coded color for each identification (see Table 5-1).

Higher ratio boosters are available on special order, but these usually require approval for purchase from the manufacturer's engineering department. It should be noted that each horn has a limit to which its amplitude can be increased without fracturing the horn.

Figure 5.8 is a graph showing how the amplitude of a typical horn can be changed by using booster horns. Figure 5.8 further illustrates how a set of three amplitude-increasing boosters can change the pressure requirements of a typical exponential horn and power-supply combination. With lower amplitudes there is a greater force capability. It is relatively easy to "stall" a high-amplitude horn by

Table 5.1
Booster Horn Codings

Amplitude Increasing		*Coupling Bar**		*Amplitude Decreasing*	
Ratio	Color	Ratio	Color	Ratio	Color
1 to 1.5	gold	1 to 1	green	1 to 0.6	purple
1 to 2.0	silver			1 to 0.5	blue
1 to 2.5	black			1 to 0.4	red

*The coupling bar is not an amplitude altering device. It is attached between horn and converter to achieve rigidity in mounting.

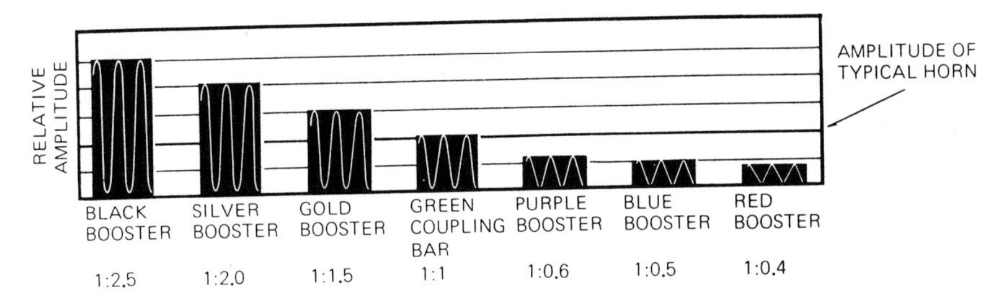

Figure 5.7. Amplitude ratios.

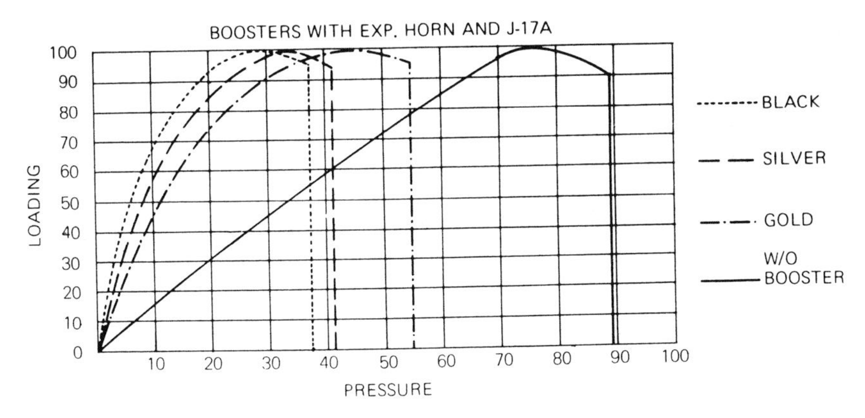

Figure 5.8. Boosters with exponential horn and J-17A.

operating it under high-pressure conditions, just as it would be easy to stall a car motor starting up a steep hill in third gear. Each horn-booster combination must be tailored to the specific application for optimum performance.

The conditions that suggest the need for altering the amplitude of a horn are listed below:

Increase Amplitude When:

1. There is difficulty getting energy to joint resulting in a poor or slow weld.
2. Energy is passing through joint (vibration can be felt in nested part; part may show marking from nest).
3. There is difficulty getting proper loading, or pressure required is beyond range of stand.

4. Diaphragming occurs. (Burnout of circular parts.)
5. If staking, melt occurs at base of stud instead of at surface.
6. Marking of parts occurs because of excessively long weld times.

Decrease Amplitude When:
1. System will not start or starts with difficulty.
2. System stalls with low pressure.
3. Excessive no-load readings occur.
4. Going from solid to tapped horn.
5. Marking of parts occurs. Higher pressure provides better coupling of vibrations into plastic.
6. Plastic parts are shattered or metal insert fractures.
7. Excessive heat builds up near nodal area in horn.
8. Diaphragming occurs.

Typical Applications

Metallurgical Applications

1. Metal deformation with reduction in cracking, i.e. riveting, swaging, and spinning.
2. Wire drawing.
3. Sintering of powdered metals.
4. Cavitation erosion studies.
5. Improving adhesional characteristics of metallics and non-metallics to accept solder, brazing, and welding alloys.
6. Stress relieving and grain modification of ferrous and non-ferrous alloys.
7. Modifying the structure and distribution of crystalline spherulites in polymers and metal alloys.

Physical Applications of Ultrasonics

1. Drilling and machining ceramics, glass, precious stones, and metals.
2. Swaging, embossing, inserting, staking, and welding of plastics.
3. Reactivation of adhesives and epoxy resins.
4. Close proximity cleaning of dies, spinnerettes, etc.

5. Extraction of alkaloids, oils, animal fats, etc.
6. Emulsification of chemicals.
7. Homogenization and disaggregation of materials.
8. Pigment dispersion of paints, dyes, etc.
9. Improvement of the flow characteristics of thizotropic materials.
10. Degassing and regassing of liquids.

Chemical Applications of Ultrasonics

1. Emulsification of chemicals.
2. Catalytic effects in chemical reactions.
3. Polymerization and depolymerization, i.e. styrene monomers.

Chapter 6

Ultrasonic Welding of Plastics

The welding of molded plastic parts by ultrasonics is accomplished with the use of a metal "tool" or "horn" which is vibrating at a rate of 20,000 cycles per second or more. These vibrations are transmitted into the plastic part by clamping it against the tool under controlled pressure for a pre-set period of time. The vibrations are transmitted into the plastic and travel uninterrupted until they meet the surface, where one part joins another. Here frictional heat is generated in the form of a complex of shear and compressional waves to melt and bond the joining surfaces.

If a plastic molded part is ultrasonically weldable, the bond is usually accomplished in one of two ways, by contact sealing (Figure 6.1A), remote sealing (Figure 6.1B), or a variation combining both methods (Figure 6.1C).

Practically all thermoplastic materials can be contact sealed, and many materials having rigid shapes may be sealed remotely. It is obvious that the energy requirements to effect fusion become less as the tool contact with the part approaches the sealing area. It is therefore wise to design the part so that the seal area will be as accessible as possible to the tool to be utilized, to prevent energy losses within the part itself.

If the contact area at the seal is reduced, the amount of energy available there becomes concentrated, producing an immediate fusion. One way of concentrating energy is by converging it to a point. This can be accomplished by molding a V-bead on the part (Figure 6.2). A good dimensional configuration for this tongue and groove is shown in Figure 6.3. The proportions shown, however, are limited to parts having wall thickness of between 0.025 in. to 0.125 in.

The wall thickness is an important consideration when designing a part for remote sealing utilizing ultrasonics. The tool is brought in contact with the part under sufficient pressure to couple vibrational energy into the plastic part itself. If the wall thickness is inadequate, the parts will deflect under this pressure

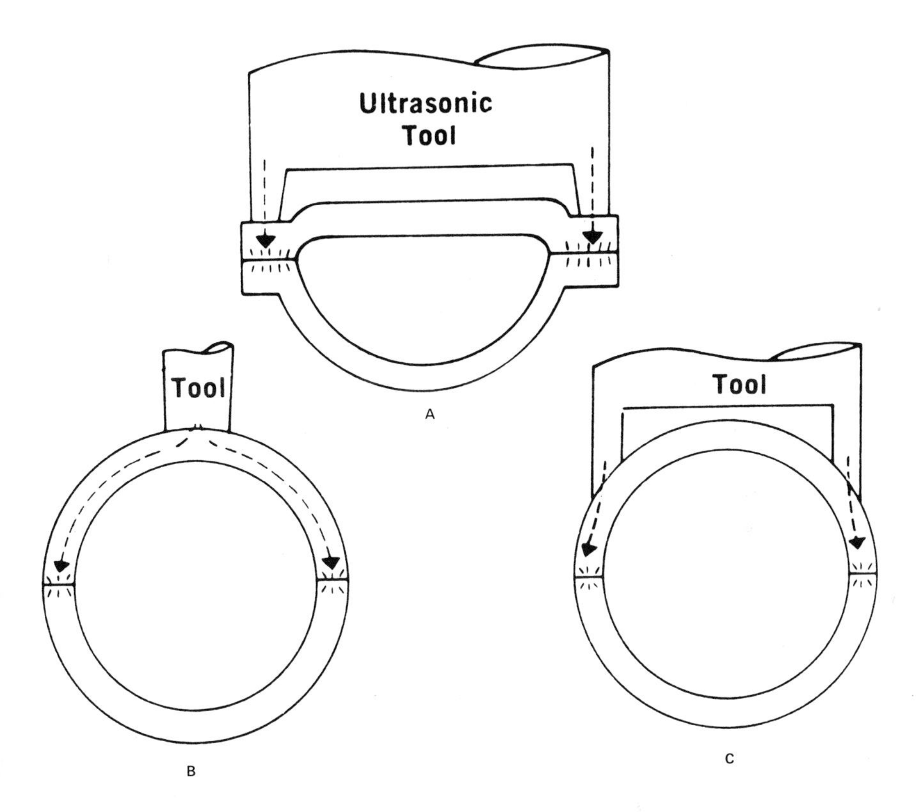

Figure 6.1. The ultrasonic tool may be directly aligned for contact sealing (A); indirectly aligned for remote sealing (B); or partially aligned to combine both methods (C).

and cause a mismatch at the joint. Parts having thin walls can sometimes be bonded if a mold of approximately half the size of the piece, usually fabricated from epoxy, is used to contain it and prevent it from distorting. Thickness of wall is desirable because it assures better transmission of energy through the part, although excessive material can cause absorption and loss of efficiency. The amount of pressure required is determined by the type of material, the size and shape of the part, as well as the availability of sufficient vibrational amplitude at the tool. Considering these variables, it is understandable that no general rule of thumb can be applied, and each problem must be evaluated individually.

The use of locating pins instead of tongue and groove is perfectly permis-

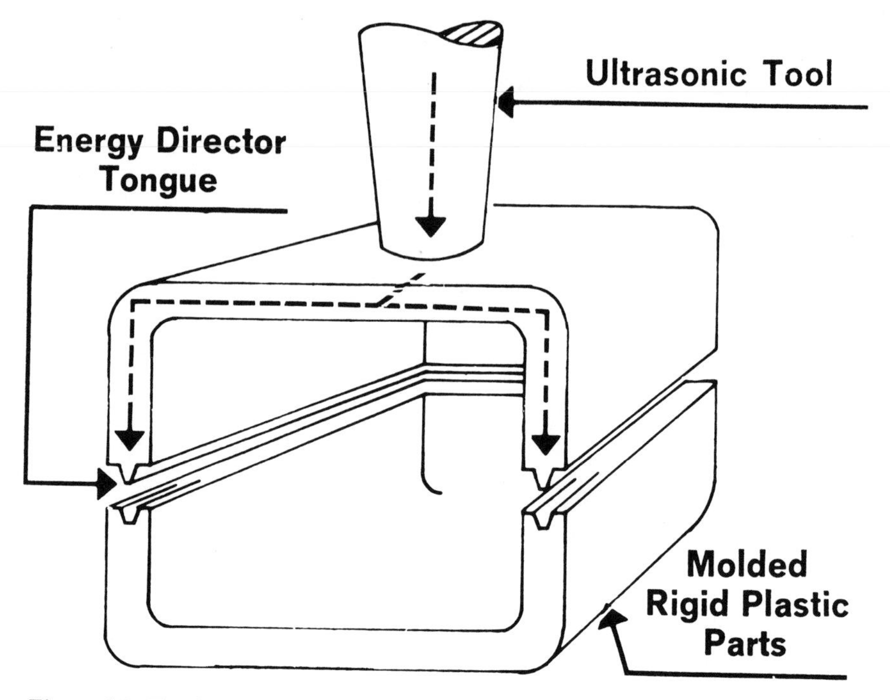

Figure 6.2. The fusion rate can be increased by concentrating the energy through a V-bead, or tongue.

sible in pieces where the tightness of the joint is not critical. The pin is usually designed to be slightly longer than the depth of the hole (0.010 in. to 0.020 in.) and about 0.010 in. to 0.015 in. smaller in diameter to allow room for material flash (Figure 6.4). After the pins fuse, and the part sinks down flush to the mating surfaces, some further spotty sealing will occur, usually where some intimate contact of the faces occurs.

Ultrasonic Plastics Welding Equipment

Various other ultrasonic machinery manufacturers produce a unit similar to the standard 400 series (Bronson), but assign different identification codes.

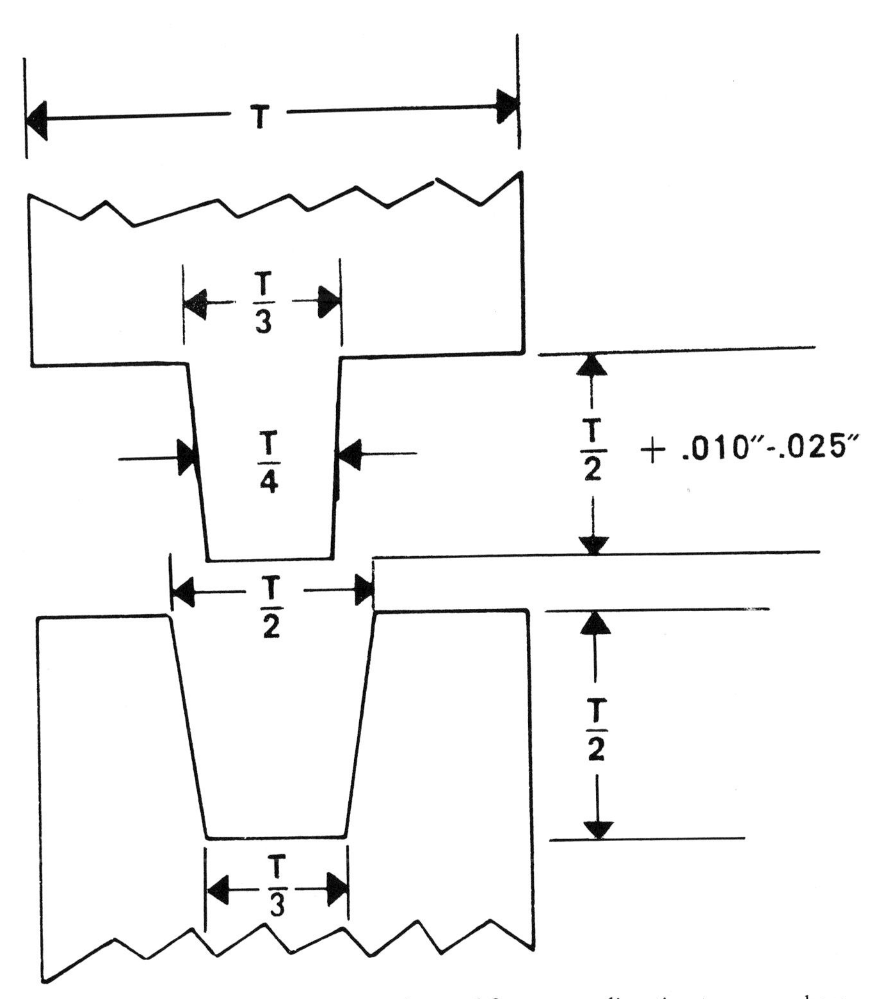

Figure 6.3. These dimensions can be used for energy-directing tongues where thickness (T) is in the 0.025–0.125 in. range.

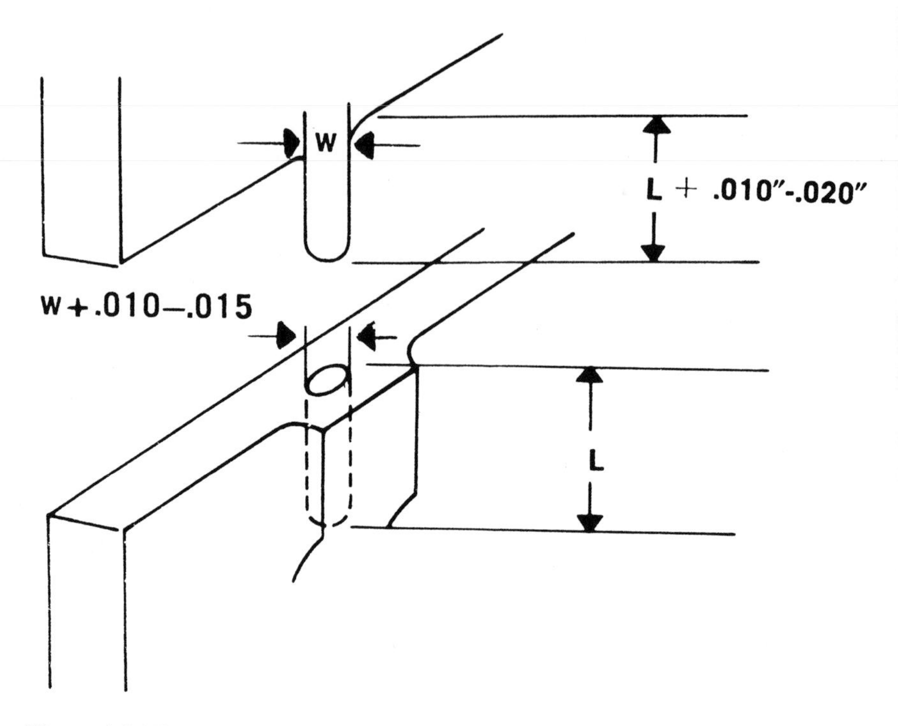

Figure 6.4. These dimensions for locating pins and holes allow excess material in the pin length to flash on fusing.

Construction

Rigidity of the 400 Series is ensured by a 3½-in. diameter, chrome-plated support column with a ½-in. wall for consistently accurate contact with parts. Specially designed clamps maintain trueness of column to base. The cast aluminum base provides a large, rugged work surface which measures 10½ in. X 14¼ in. The pneumatic carriage and power supply housings are also made of cast aluminum. Each is modular in construction for ease of replacement and servicing.

Operation

The 400 Series is an integral power supply/stand system, requiring only two external connections–line voltage and air supply. Converters are easily snapped in and out, with no connecting cables, through a swing-open access panel. Converters are free to rotate 360 degrees to ease alignment with parts to be assembled with the horn. Time cycle and pressure controls are mounted on the front panel, as well as an activity meter and test switch for monitoring power delivery.

Versatility

The 400 Series is capable of welding, staking, inserting, swaging, and spot welding thermoplastic parts. The 3-in. stroke of the reciprocating carriage accommodates parts with deep cavities (a 6-in. stroke is available on special order). Independent carriage down-speed and return-speed adjustments, with dynamic trigger pressures of 10 to 200 pounds, offer greater applications range. The head can be rotated about the column and can be positioned at angles to the base. The head can also be mounted on flat surfaces, such as I-beams, which allows multi-head setups for special systems. Height is easily adjusted by a crank on the side of the head. The crank can be removed to allow several systems to be stacked side by side as close as 6¼ in., center to center.

Safety

The 400 Series features non-locking palm buttons on the base panel to keep the operator's hands away from the work area during operation. An emergency stop button is also provided on the base panel. A counterbalanced elevation control eases head adjustment on the column and prevents the head from dropping unexpectedly during the adjustment. All models of the 400 Series welders are FCC approved.

Specifications

Power Requirements: 117 VAC, 60 Hz, 8 amps.
Air Supply: Clean, dry, dirt- and oil-free air supply
Down Pressure on Part: 480 lbs. @ 100 psig
Dynamic Pressure Trigger Range: 10 to 200 lbs.
Output Power: Model A – 2900 in.-lbs./sec. mechanical power to plastic;
350 watts electrical power to converter
Model B – 5900 in.-lbs./sec. mechanical power to plastic;
700 watts electrical power to converter

Output Frequency: 20 kHz (20,000 cycles per second)
Weld Cycle Range: 0.1 - 6.0 seconds
Hold Cycle Range: 0.05 - 3.0 seconds
Dimensions: Base: Height – 3¾ in.
Width – 16¼ in.
Depth – 22⅞ in.

Column: Diameter – 3½ in.
Length – 36 in. (longer columns available on special order)
Wall thickness – ½ in.

Housing: Height – 27 in.
Width – 5¾ in.
Depth – 13¾ in.

Stroke: 3 in. (6-in. stroke available on special order)

Throat: 12⅜ in. (center of horn to center of column)

Weight: 160 lbs. (includes power supply, stand, base and column)

Spot Welding

Spot welding thermoplastics in a split second with an ultrasonic pistol-grip hand tool is now possible for non-stationary requirements. Compact, lightweight and portable, the tool spot welds large parts, and those with hard-to-reach joining surfaces, with ease, in your plant or in the field. The spot welds are good looking and strong in both shear and peel.

Applications. Spot welding may be used to assemble curtain wall panels, appliances, aircraft ductwork, furniture, car bodies, snowmobiles, amphibious vehicles, siding for houses, gutters, downspouts, animal shelters, toolsheds, trailer bodies, storage facilities, and for any other application that requires the assembly of large thermoplastic parts. Materials may be ABS, PVC, polypropylene or almost any other thermoplastic.

Operating Techniques. Vibrating ultrasonically, the pilot of the tip passes through the top sheet and enters the bottom sheet to a depth of one-half the top sheet thickness. The molten plastic displaced is shaped by a radial cavity in the tip and forms a neat raised ring on the surface. Simultaneously, energy is released at the interface, producing essential frictional heat. Molten plastic displaced from the second sheet flows into the preheated area and forms a permanent molecular bond.

Standard threaded tips are available for spot welding materials ranging in thickness from 0.032 in. through 0.218 in. Larger tips are available for stock gauge sheets up to 0.312 in. on special order.

	Thickness of Top Sheet	
	1/32 in.	0.032 in.
	3/64	.047
Threaded for	1/16	.062
½-in Tapped Horn	5/16	.078
	3/32	.093
	7/64	.109
	1/8	.125
Threaded for	5/32	.156
¾-in Tapped Horn	3/16	.187
	7/32	.218

Accessories. The compression collar controls the depth of penetration and presses the two sheets together so that a consistent weld is achieved. Available in ½-in. and ¾-in. diameter sizes. The compression collar includes a jacket fitted for air cooling and a special fluted horn to dissipate heat generated by continuous production use.

Typical Specification.

Length including horn:	13¼ in.
Width including grip:	8½ in.
Depth:	3¼ in.
Weight:	4½ lbs. (without cable)
Power suppiies:	System A and System B
Power output:	System A – 1700 in.-lbs./sec.
	System B – 3900 in.-lbs./sec.

Principle of Assembly. Sixty cycles per second electrical energy is converted to 20,000 cps electrical energy by a power supply. A converter then transforms the electrical energy into mechanical vibratory energy at 20,000 cps. The mechanical vibratory energy is transmitted to a tool or horn. When the tool contacts the part with high-frequency vibrations, friction—and thus localized heat—is created between the tool and the top sheet and between the two sheets. The plastic melts, flows and forms a permanent bond in a split second.

Scan Welding

Large, flat, rigid thermoplastic assemblies can be economically welded with scan welders. This recent development makes it possible to weld moving rigid parts with one or more horns. Scan welding offers maximum welding rates with minimum equipment investment.

Applications. Scan welding may be used to assemble thermoplastic panels and doors for appliances, furniture and cabinets as well as other large parts that have large flat surfaces. There is no limit to the length of a scan welded part. Any width may be accommodated with the addition of more welding heads.

Principle of Assembly. Sixty Hz electrical energy is converted to 20,000 Hz electrical energy by a power supply. A converter then transforms the electrical energy into mechanical vibratory energy at 20,000 Hz. The mechanical vibratory energy is transmitted to the horn. When the horn contacts the part with high-frequency vibrations, this vibratory energy passes to the joint area where friction creates localized heat between the two parts. The plastic melts, flows and forms a permanent bond.

Automation Capability. Scan welders are available with either slide or continuous belt. The continuous belt unit is easily incorporated into an existing conveyor system and will handle a greater range of part sizes.

Gang Welding

An ultrasonic gang-welding system consists of an ultrasonic power supply which simultaneously triggers up to eight gang-welding units (heads). Each head is driven by its own power module contained in the system's power supply.

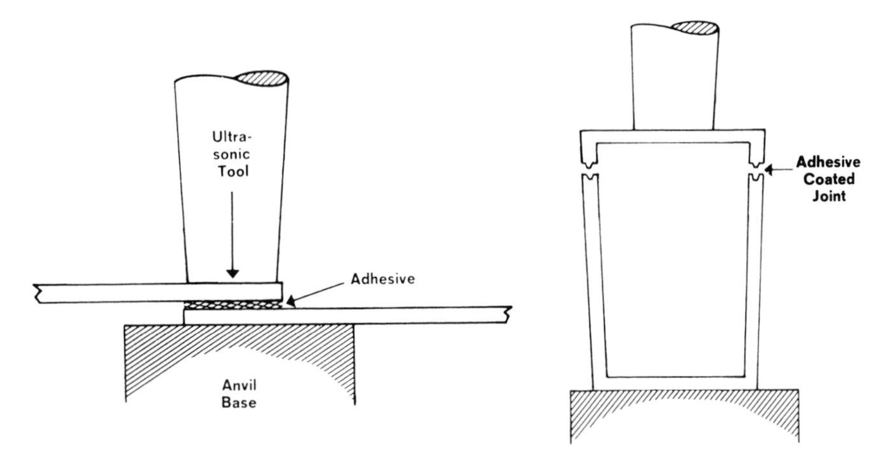

Figure 6.5. Energy direction designed joint.

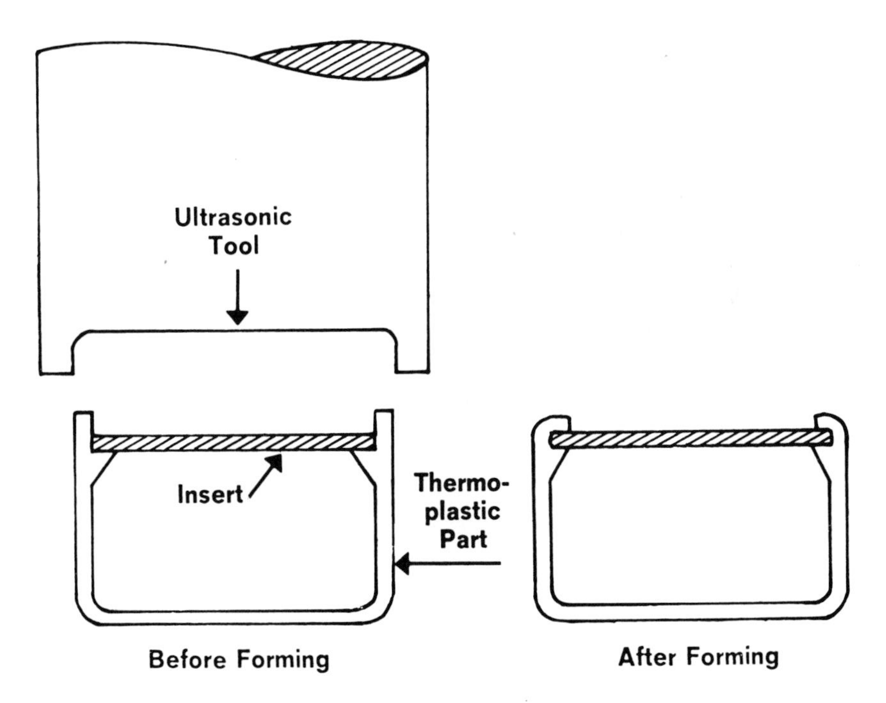

Figure 6.6. Example ultrasonic insert forming.

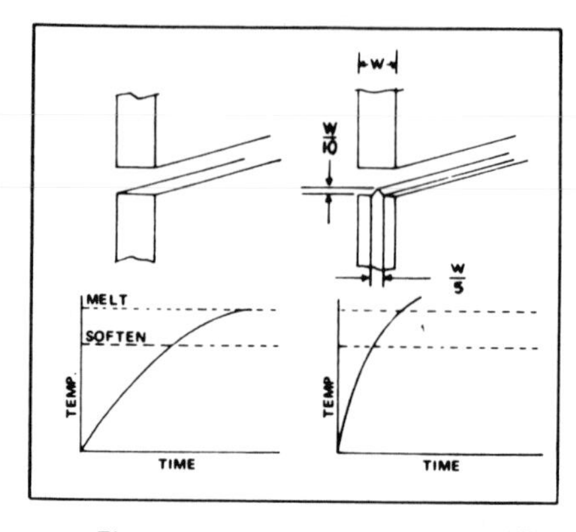

Time-temperature curves compare welding of a butt joint with and without an energy director. Width ratios show energy director recommended proportions.

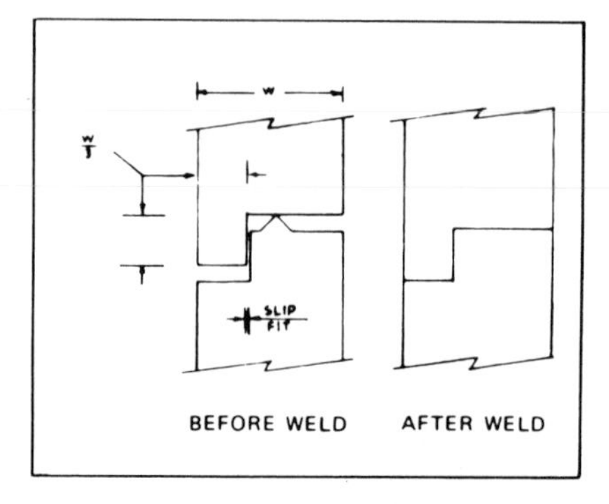

Step joint (top) is useful where a weld bead on the side would be objectionable. Tongue-and-groove joint (bottom) usually has the capability for greatest strength.

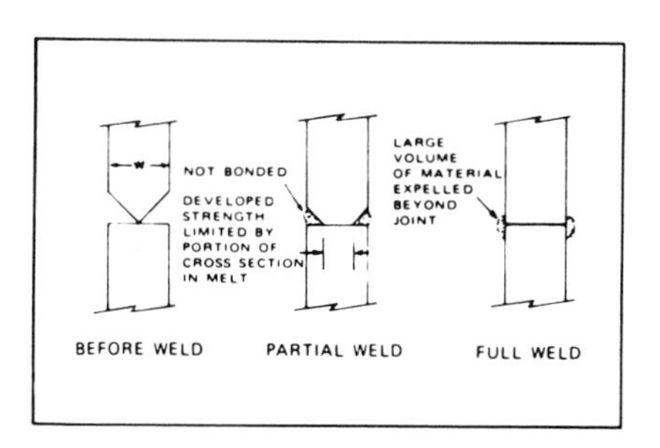

A typical mistake is beveling one joint face at 45°, instead of using an energy director. Either a weak or sloppy joint will result if an energy director is not used.

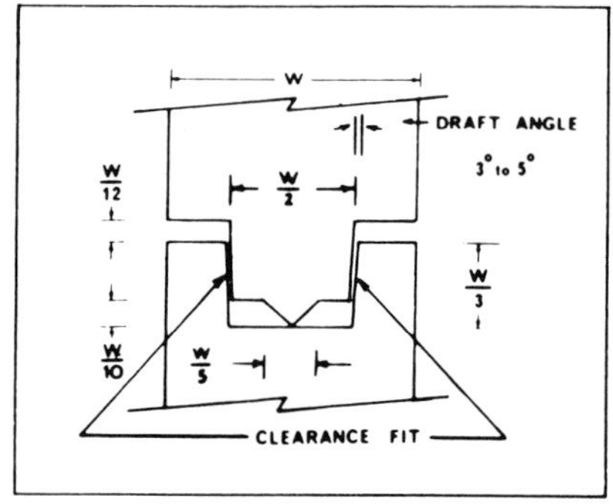

Figure 6.7. Butt joint designs.

The number of welding heads is determined by the number of welds required to assemble the part. A single system can power up to as many as eight heads. To accommodate the contours of larger parts, any number of units can be slaved together to drive an unlimited number of heads, all activated simultaneously.

Individual welding heads and modular power supplies simplify system design, and heads are easily repositioned to allow in-plant retooling. Converter mountings and wiring are ruggedly constructed; cable lengths and connectors are not acoustically critical. Since the system allows tuning of individual heads, each horn tip in a gang-welding system can be of a different configuration. And each power module can be monitored separately to ensure optimum energy delivery.

Gang-Welding Head. Each gang-welding head consists of a converter mounting plate, a pivoting anvil operated by an air cylinder, and a ball shaft mounting base. The plate can be adjusted vertically on the ball shaft to a maximum height of 2 in. for standard clamps, although longer shafts are available. The plate is secured to the ball shaft by a clamp which allows 360-degree rotation of the plate about the shaft. The plate can be pivoted approximately 40 degrees in a vertical plane at its mounting clamp. The entire fixture can be rotated in a 50-degree cone about the ball socket. Clamping of the part by the pivoting anvil is accomplished by a 1-in. diameter air cylinder moving through a 2½-in. range. The speed at which the anvil will clamp is controlled by individual air-flow valves. Heads can be placed as close as 2 in. on center in any of the three dimensional planes.

Power Supply. The ultrasonic power supply operates from standard 117-volt AC, 60-Hz line power. The power supply can handle up to eight converters, delivering 1300 in.-lbs./sec. of energy at 20 kHz to each tip. An aluminum case houses the solid-state, modular circuitry. Housing dimensions are 24¾ in. wide × 24¼ in. deep × 6 in. high. A front panel activity meter is used to monitor each head, as selected by a switch, during tuning or operation. Weld and hold time cycles are adjustable for all converters by one set of controls. A connection is provided to link multiple power supplies together to allow simultaneous activation of all heads by a single switch.

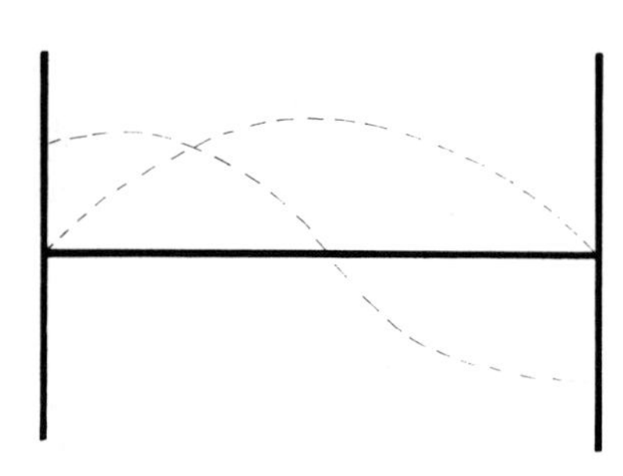

Figure 6.8. Sonic Converters and horns.
6.8A. Converters–Principles of operation: In a resonant, uniform half-wave section, the following can be observed. (1) Velocity is maximum and opposite at the ends, zero at the node; (2) stress in the material, which is alternately in tension and compression, is maximum at the node and zero at the ends.

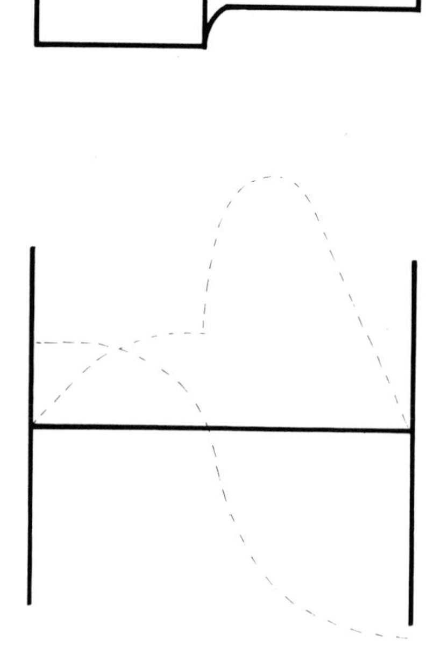

6.8B. Shapes–stepped horns: The stepped horn consists of two quarter-wave sections of different but constant cross-sections joined together at the node with a relatively small radius. This type of horn produces very high gain, but the horn material becomes very highly stressed when driven to high amplitudes.

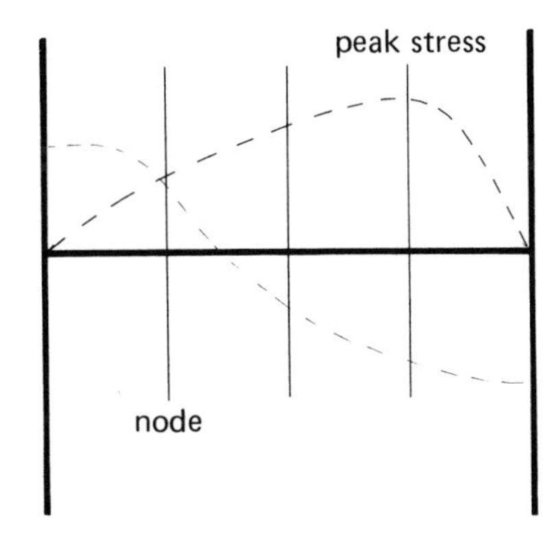

6.8C. Exponential horn: The exponential horn has a very desirable stress distribution, but because the gain is so low, it finds limited use in plastics welding. The peak stress is displaced to the right of center point of the horn by an amount equal to the displacement of the nodal (no motion) point of the horn.

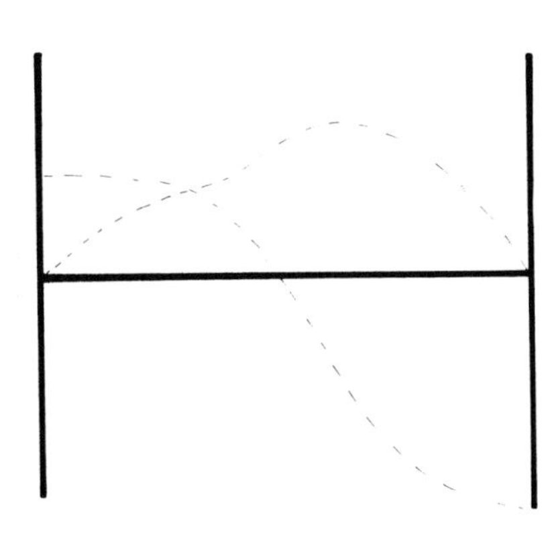

6.8D. Catenoidal: The catenoidal horn combines the desirable gain of the stepped horn with the good stress-distribution of the exponential horn. It is most suitable for plastics welding and the staking of small parts.

Chapter 7

Ultrasonic Inserting and Staking

Inserting

A new ultrasonic technique for inserting and encapsulating metal in plastic can now replace the costly and time-consuming conventional method of molding plastic around the metal at the molding machine.

A hole that is slightly smaller in diameter than the insert it is to receive is premolded in the plastic. Such a hole provides a certain degree of interference-fit and guides the insert into place. For a final interlocking assembly, the metal insert is usually knurled, undercut, or shaped to resist the loads imposed on the finished assembly.

Ultrasonic vibrations travel through the driven part until they meet the joining area between metal and plastic. At this joint (or interface) the energy of the ultrasonic vibrations is released as heat. The intensity of heat created by the vibration between the plastic and the metal is sufficient to melt the plastic momentarily, permitting the inserts to be driven into place.

Ultrasonic exposure time is usually less than one second, but during this brief contact the plastic reforms itself around knurls, flutes, undercuts, or threads to encapsulate the insert.

Example. A typical assembly consists of a knob of impact styrene and a steel insert. If a knob of this type is used as a locking device, it must withstand torque-loading as it is tightened on a threaded rod. It must also withstand axial shear forces as pressure is brought to bear against both plastic and metal insert surfaces.

Design Requirements. Insert/hole design will vary with each application, but a sufficient volume of plastic must always be displaced to fill voids created

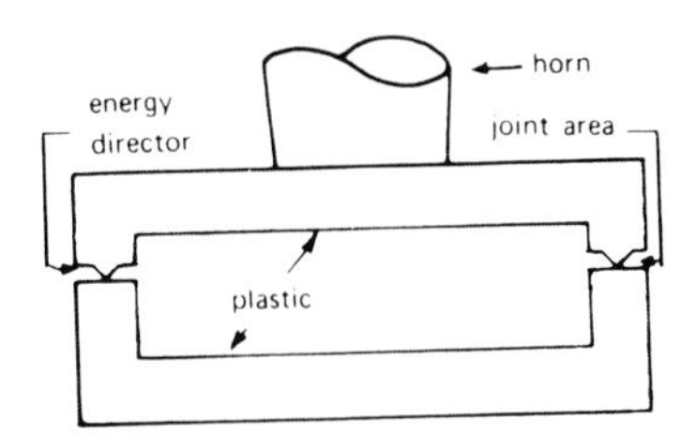

In ultrasonic welding, ultrasonic energy melts an energy director at the weld interface, joining two sections of plastic, usually in less than a second.

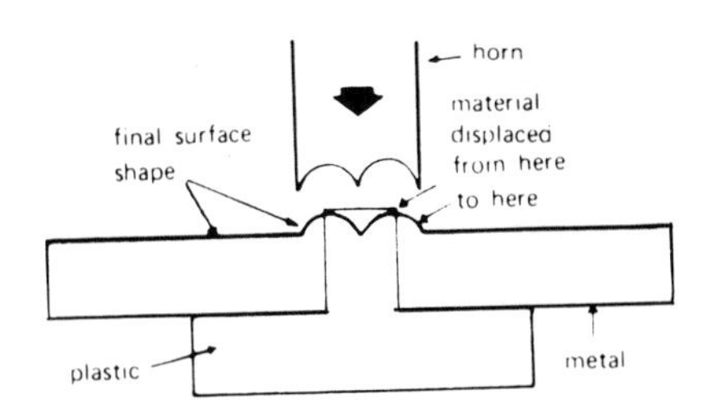

In ultrasonic inserting, a metal insert is driven with applied force and ultrasonic energy into a hole of slightly smaller diameter, melting the plastic momentarily.

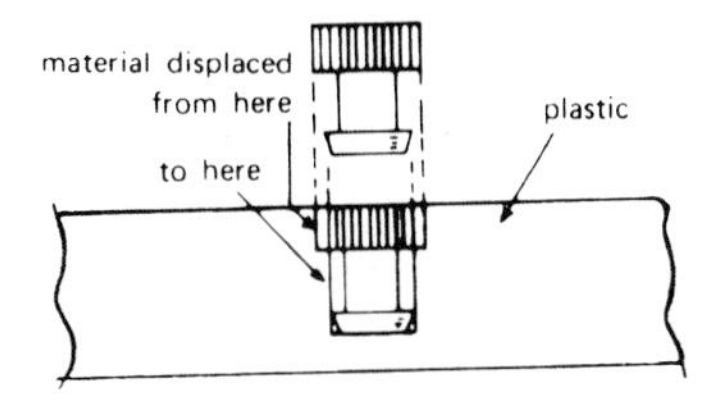

In ultrasonic staking, applied force and ultrasonic vibration distort a plastic stud, allowing a plate to retain a metal section.

Figure 7.1. Application of ultrasonic energy achieves different objectives in the welding, inserting, and staking procedures.

by knurled and undercut areas of the insert. A slight excess of molten material can usually be tolerated, while insufficient interference may result in less than required strength.

The choice of contacting the plastic or metal surface with the horn will depend upon the configuration of the part and the ability of the assembly to accept the required vibratory intensity.

Staking

In staking, the opposite is usually true. The flow properties are easier to control on soft plastics, which require less set-up and final adjustment time.

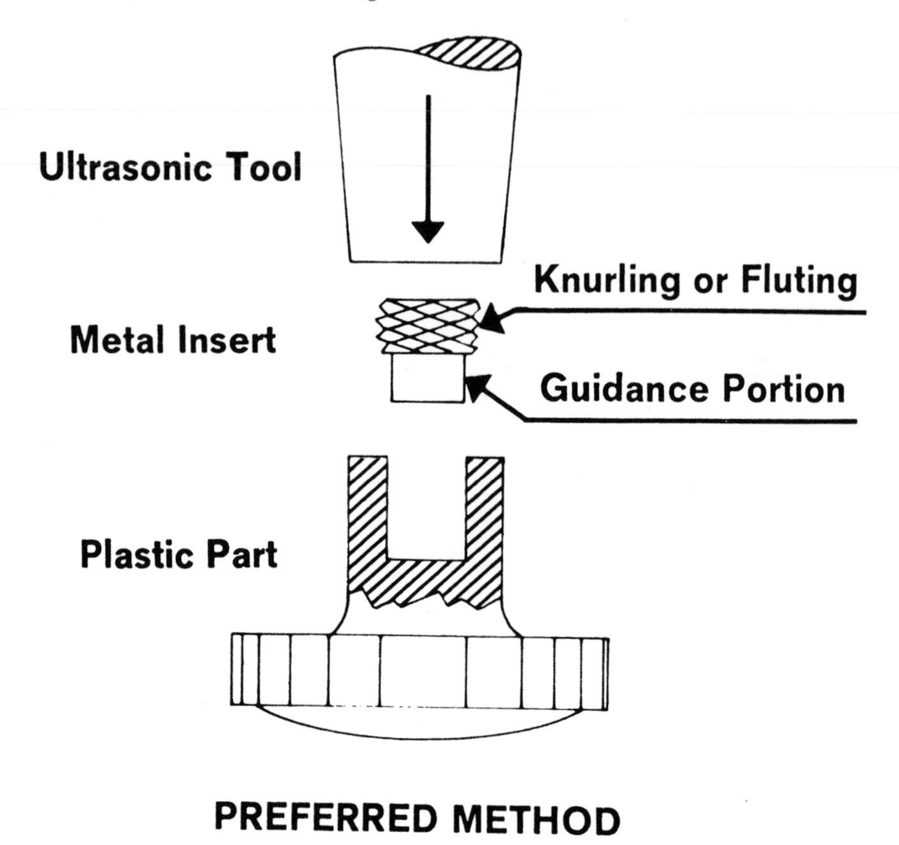

Figure 7.2. Sufficient excess plastics must be displaced to flow around and seal the insert in place.

However, good results can be achieved with most plastics when the right amplitude and force combination is used.

General. As with ultrasonic welding and inserting, ultrasonic staking employs the same principle of creating localized heat through the application of high-frequency vibrations. Most staking applications involve the assembly of metal and plastic.

Example. In staking, a hole in the metal receives a plastic stud. Ultrasonic

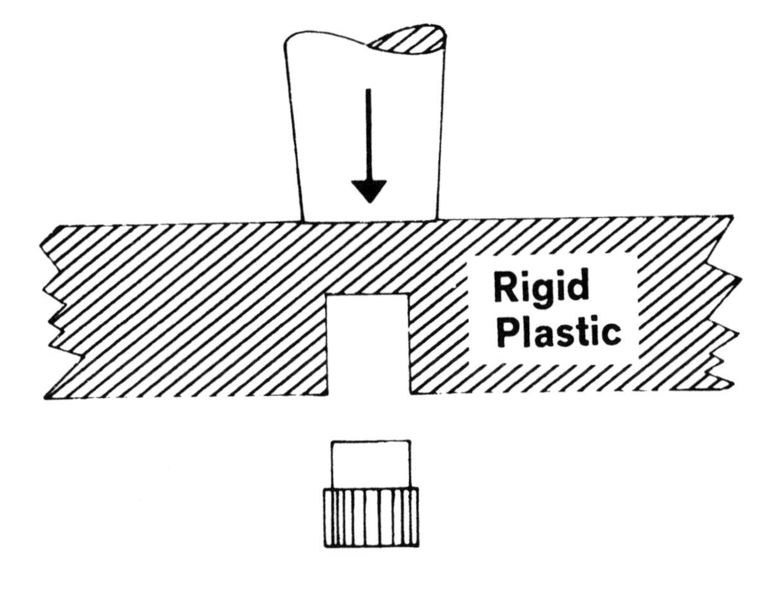

ALTERNATE METHOD

Figure 7.3. It is usually preferable to have the ultrasonic horn contact the metal insert, as shown in Figure 7.2; however, the horn may instead contact the plastics when metal contact is impossible.

staking requires the release of vibratory energy only at the surface of the plastic stud; therefore, the contact area between horn and plastic must be kept as small as possible. The horn is usually contoured to meet the specific requirements of the application. With the introduction of ultrasonic vibrations, the stud melts and reforms to create a locking head over the metal.

Design Requirements. There are two head-forms that will satisfy the requirements of a majority of applications. The first, generally considered standard, produces a head having twice the diameter of the original stud, with a height one-half the stud diameter. The second, referred to as a low-profile head, has a head diameter one and one-half times the stud diameter, with a head height one-quarter the size of the head diameter.

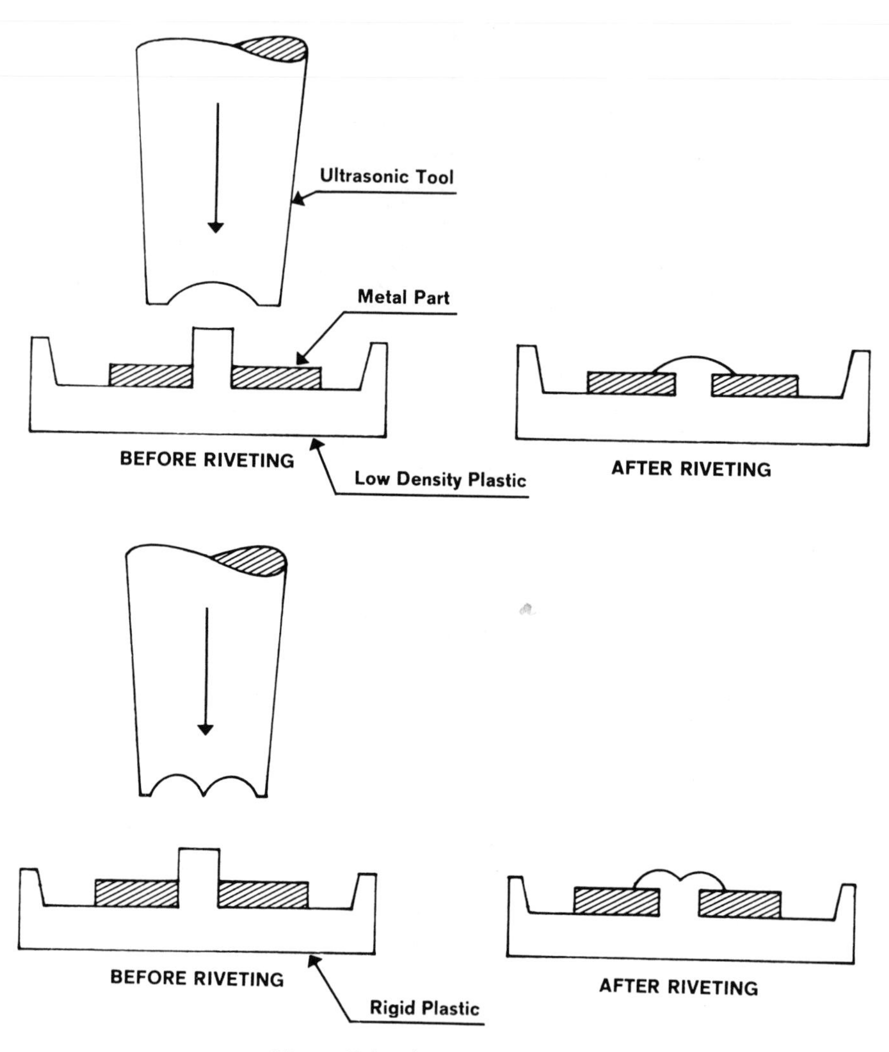

Figure 7.4. Ultrasonic riveting.

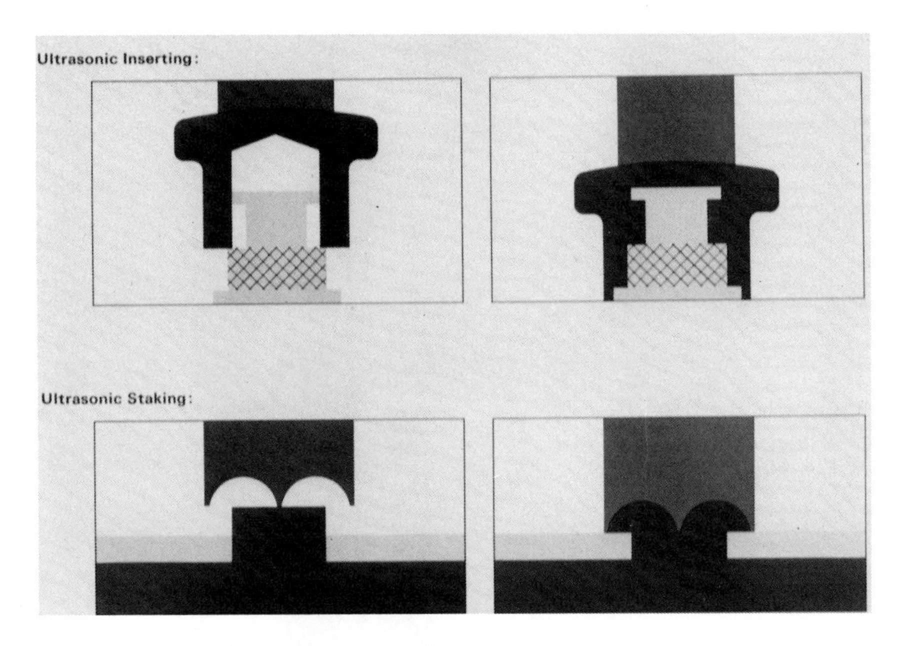

Figure 7.5. Ultrasonic inserting.

Set-up. Unlike ultrasonic plastics welding, staking requires that out-of-phase vibrations be generated between horn and plastic surfaces. Light, initial contact pressure is therefore a requirement for out-of-phase vibratory activity within the limited contact area. It is the progressive melting of plastic under continuous, but light, pressure that forms the head. Adjustment of flow-control valve and trigger switch may be required to reduce pressure to the desired level (see Table 7.1).

Table 7.1
Header Profiles for Proper Flow

Head Form	*Diam.*	*Diam.*	*Height*	*Center-Center Diam.*	*Stud Height Above Part Before Heading*
Standard	d	2d	0.5d	d	1.6d
Low Profile	d	1.5d	0.25d	0.75d	0.6d

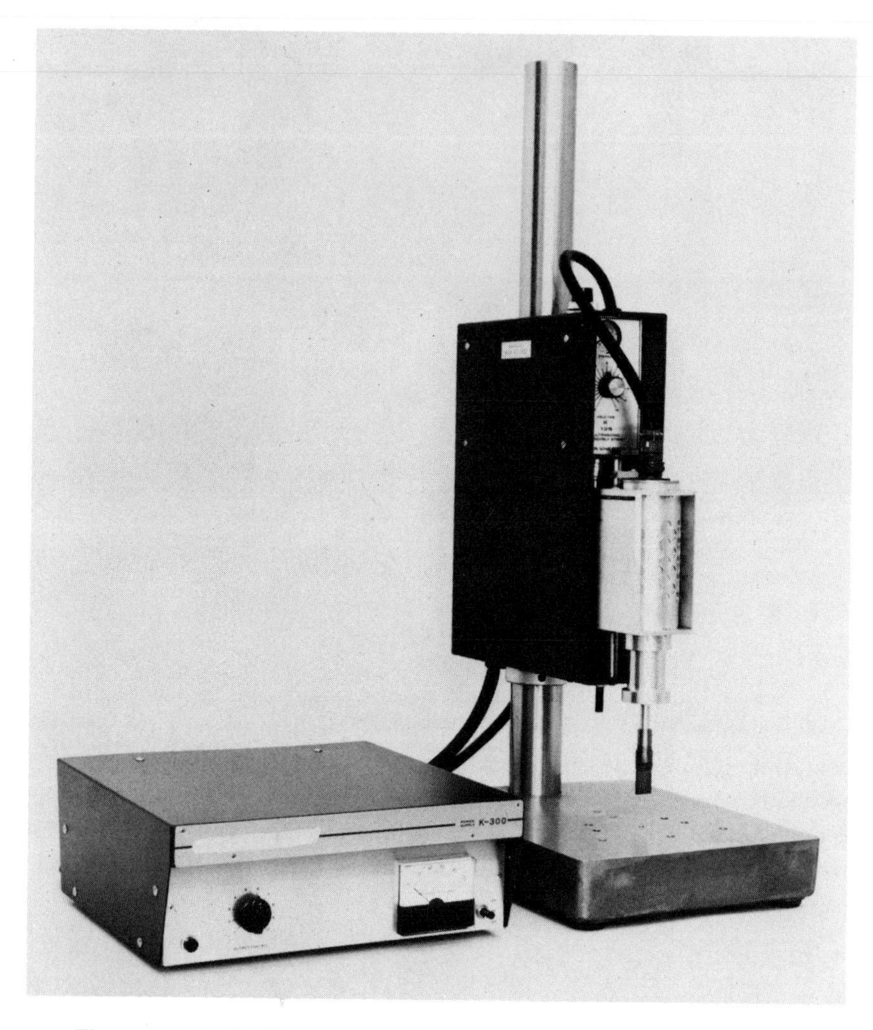

Figure 7.6. A 40 kHz system, utilized in the assembly of small parts.

Table 7.2
Characteristics of Materials in Welding, Staking and Inserting.*

Application	Material	Staking and Inserting	Welding† Near-Field	Welding† Far-Field	Remarks
General-Purpose Toys Appliances Housewares	Polystyrene unfilled	E	E	E	Excellent accoustical properties; produces strong, smooth joints.
	Rubber-modified	E	E	G - P	Welding characteristics depend on degree of impact resistance.
	Glass-filled	E	E	E	Weldable with filler content up to 30%.
	SAN	E	E	E	Particularly good as glass-filled compounds.
	ABS	E	E	G	Can be bonded to other polymers such as SAN, styrene, and acrylics.
Engineering Plastics Automotive Appliance Electronic	Polycarbonate	E	E	E	High melting temperature requires high energy levels.
	Nylon	E	G	F	Oven-dried or "as molded" parts perform best due to hygroscopic nature of the material.
	Polysulfone	E	G	G - F	
	Acetal	E	G	G	Requires long, high-energy ultrasonic exposure because of low coefficient of friction
	Acrylics	E	E	G	Weldable to ABS and SAN; applications include dials, radio cases, and meter housings. In sheet form, joints must be machined.
	Polyphenylene oxide	G	G	G - F	High melting temperatures require high energy levels
	Noryl	E	G	E - G	
	Phenoxy	E	G	G - F	
High-Volume Low-Cost Applications	Polypropylene (Olefins)	E	G - P	F - P	Horn design for welding is particularly critical; filled compounds usually better, but need individual testing.
	Polyethylene (Olefins)	E	G - P	F - P	
	Butyrates	G - F	P	P	Weldability varies with formulation and part configuration, however, these materials usually perform well in staking and inserting applications. Decomposition of some formulations may occur.
	Cellulosics	G - F	P	P	
	Acetates	G - F	P	P	
	Vinyls	E - F	F - P	F - P	

*E = Excellent; G = Good; F = Fair; P = Poor.

†Near-field welding refers to joint ¼-in. or less from area of horn contact; far-field welding to joint more than ¼-in. from contact area.

Section III

Applications

Chapter 8

Assembly, Fastening, and Joining

Ultrasonic assembly applications have multiplied in the automotive, electronic, small appliance, cosmetic, toy, housewares, and furniture industries. A relatively unknown fastening technique only a few years ago, the system today is being used in a growing number of applications, ranging from insertion of a metal body into a plastic lipstick tube to welding two halves of a plastic auto body shell.

In ultrasonic assembly, material is joined extremely fast, usually in a fraction of a second, through a combination of pressure and mechanical vibrations at an ultrasonic frequency. The process is timed precisely for each application to ensure optimum joining.

Widespread adoption of ultrasonic assembly methods probably is due to the numerous advantages that the system offers. In general, the assemblies are reliable, neat appearing, and can be made rapidly.

Systems for ultrasonic plastics assembly have evolved into a number of sizes and styles to satisfy the increasing applications.

Assembly Systems for Joining Plastics

Assembly systems for ultrasonic joining of plastics are of two types: fixed and portable. A fixed system consists of power supply, sonic converter, horn and stand. A portable system consists of power supply and a combination sonic converter-horn equipped with a grip and on-off switch so it can be moved readily to a work station.

The power supply changes the 60-cps line voltage to 20 kHz. The power supply is available with various power output ratings. Presently, power supplies

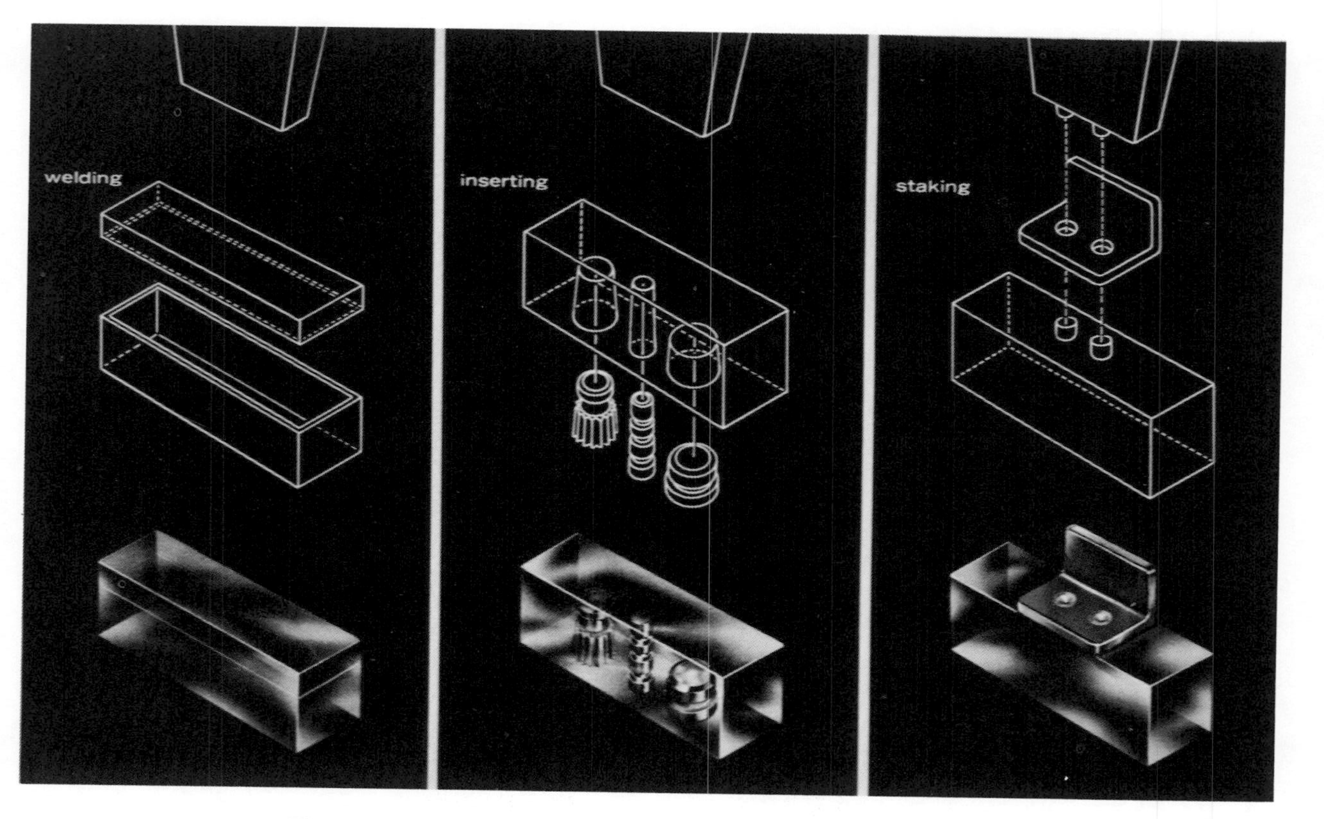

Figure 8.1. The techniques of welding, inserting, and staking are the basic procedures used in assembling plastics components.

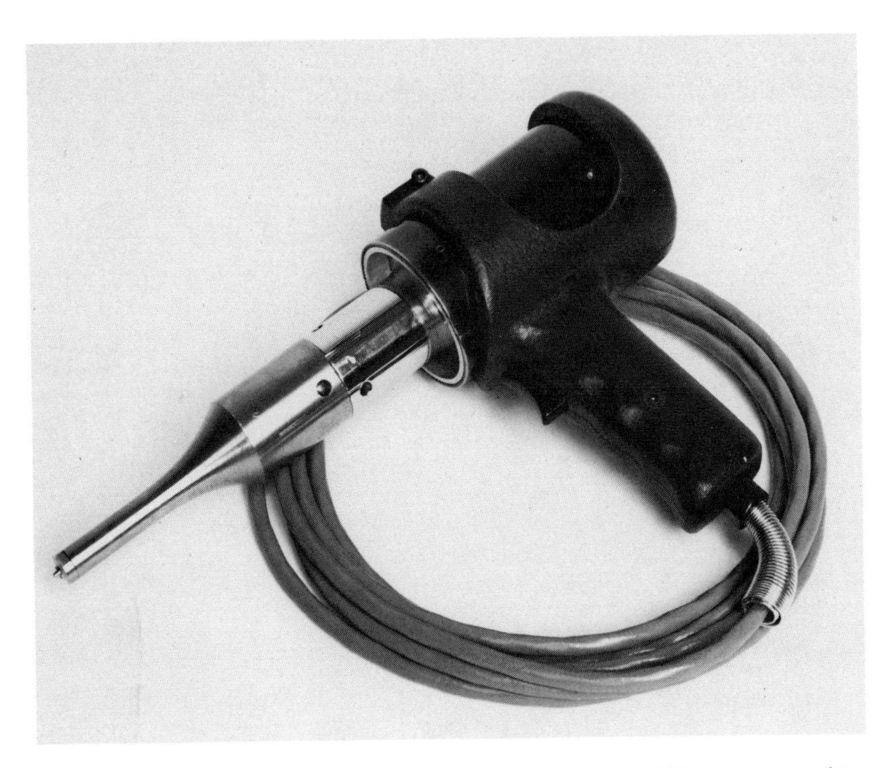

Figure 8.2. Compact, lightweight portable unit can spot-weld large parts and is convenient for hard-to-reach locations.

are marketed in the 10, 20, 25 and 40 kHz ranges. The application to be performed is the primary determinant as to selection.

The sonic converter changes the 20 kHz electrical energy into mechanical oscillations at the same frequency. There are two methods of conversion currently adopted by the industry: magnetostrictive and piezoelectric. The magnetostrictive method employs a transducer stack composed of magnetostrictive elements such as nickel, vanadium, Permendur, or ferrite. The piezoelectric method does the conversion by means of a piezoelectric element made of lead zirconate titanate. Converter efficiencies range from 50% to 90%. Available mechanical output power ranges from 300 to 35,000 in.-lbs./sec.

The horn is a half-wave resonant metal section designed to transmit mechanical vibrations at an amplitude of from 0.0005 to 0.005 in. from the

Table 8.1
Characteristics of Ultrasonic Assembly Equipment.

Power Supplies			
System	*Power output at face of horn*	*Weight*	*Dimensions*
A	2200 ips*	25 lbs.	Height 6-1/4 in. Width 16-1/2 in. Depth 18-1/2 in.
B	4400 ips	26 lbs.	Height 6-1/4 in. Width 16-1/2 in. Depth 18-1/2 in.
C	8800 ips	46 lbs.	Height 6-3/4 in. Width 20-3/4 in. Depth 24 in.

*Inch pounds per second.

Ultrasonic Assembly Stands			
Stand	*Pressure on part*	*Weight*	*Dimensions*
A	125 lbs. at 100 psig†	60 lbs.	Base Height 2-5/8 in. Width 12 in. Depth 15-5/8 in. Column Height above base 33-1/2 in. Diameter 2-1/2 in. Housing Height 16 in. Width 3-5/8 in Depth 16 in.
B	160 lbs. at 125 psig	75 lbs.	Base Height 2-5/8 in. Width 12 in. Depth 15-5/8 in. Column Height above base 33-1/2 in. Diameter 2-1/2 in. Housing Height 20-1/8 in. Width 3-3/8 in. Depth 17 in.
C	320 lbs. at 125 psig	165 lbs.	Base Height 4 in. Width 15 in. Depth 22 in. Column Height above base 32 in. Diameter 3 in. Housing Height 22-3/4 in. Width 6-1/4 in. Depth 16 in.

†Pounds per square inch gauge.

sonic converter to the parts being assembled. Each horn is designed to meet the requirements of a specific application. The requirement to oscillate at 20 kHz influences the horn mass and shape. For this reason, no two horns are exactly alike. Horns are usually constructed of a special titanium alloy to provide a high strength-to-weight ratio to withstand intense oscillating stresses and to transmit vibrations efficiently. Lower-amplitude horns can be constructed of aluminum, which offers excellent acoustical transmittal but only one-third the tensile strength of titanium alloy.

Each horn is designed to resonate vertically on each side of a horizontal node line. Horns can be built in a broad range of shapes and sizes; however, horns having certain critical dimensions must be engineered carefully to avoid the tendency to resonate in a nonvertical node, taking away energy from the desired vertical oscillation. Successful horns have been built as large as 12 in. in diameter in a slotted cylindrical configuration and 10 in. by 1 in. in a rectangular configuration. The five basic horn shapes are: exponential, step, catenoidal, rectangular and cylindrical.

Booster horns are available to increase or decrease horn amplitude to produce the proper degree of melt or flow in the plastic part. The proper horn amplitude is determined by the type of plastic, part shape, and the nature of the work to be performed. Boosters can provide for an increase or decrease of amplitude by as much as 400%. Each horn has a limit to which its amplitude can be increased without damaging the horn.

The stand consists of a base and column on which is mounted the operating head, and it supports the horn/converter assembly. Front-panel operating controls, along with the necessary control elements, include a pressure regulator and programmer adjustment controls for weld and hold time.

Weld time is the interval during which the horn, under a given pressure, vibrates at the ultrasonic frequency against the workpiece. Hold time is the interval following weld time that the horn is retained against the workpiece to allow hardening of the melt to ensure a secure joint before the horn is removed. In many applications, combined weld and hold times are under 1 second.

Other stand features include speed controls, air filter and oiler, and height adjustment. The stands are available in various sizes for different applications.

In portable systems, the sonic converter-horn assembly can be moved readily to the work site. Several versions, including a pistol grip, are available.

The systems, both fixed and portable, can be used for welding, seaming, inserting, staking, and spot welding.

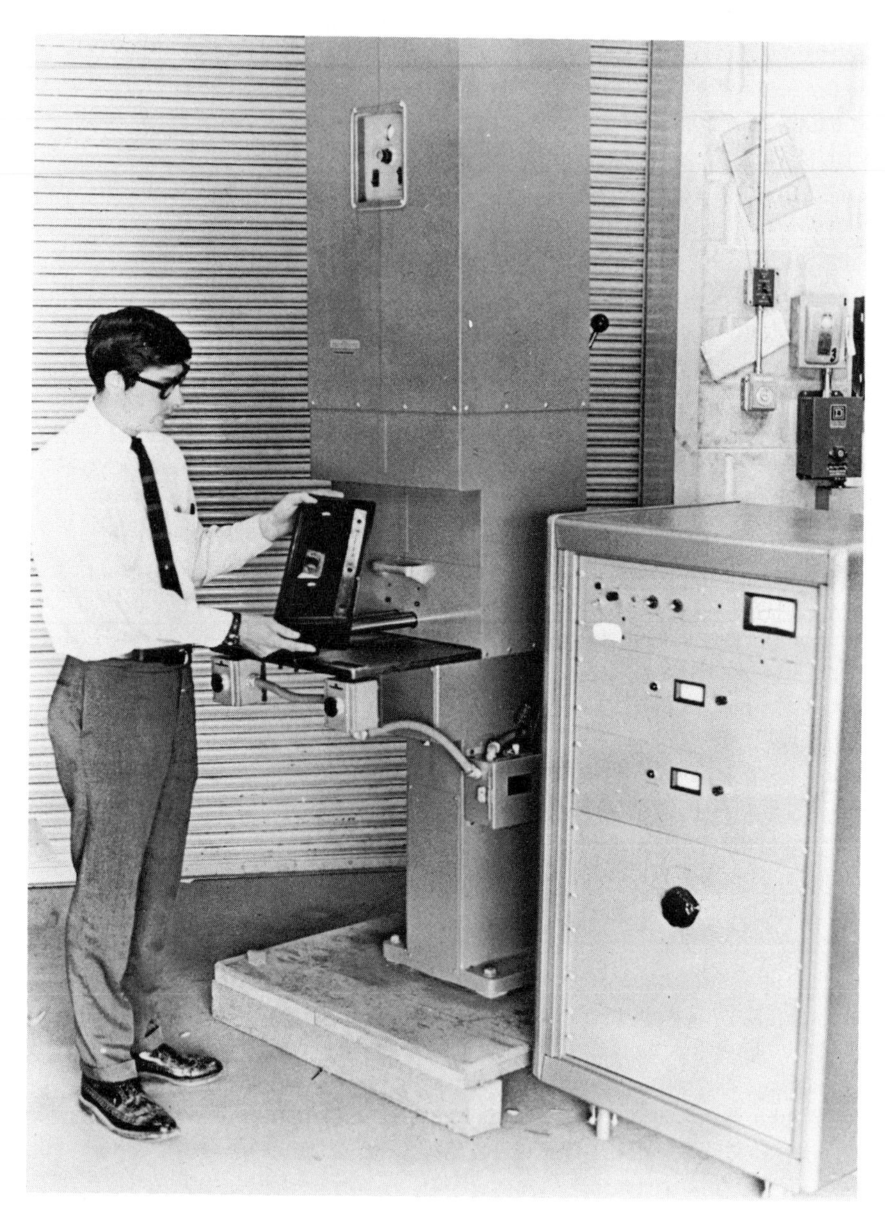

Figure 8.3. A 10 kHz plastics welder. Up to 4000 watts are available, and parts as long as 12 in. can be welded in a single operation.

Welding and Seaming

In welding and seaming of plastics, the horn is coupled to the upper part to be welded so that part vibrates against the lower part at the ultrasonic rate to produce friction and heat at the interface of the two parts.

Most thermoplastics can be assembled ultrasonically without the use of solvent, heat or adhesives. The process is extremely fast and results in strong and clean joints.

Weldability of plastics depends on melt temperature, modulus of elasticity, impact resistance, coefficient of friction and thermal conductivity. Generally, the more rigid the plastic, the easier it is to weld. Low modulus materials, such as polyethylene and polypropylene, often can be welded provided the horn can be positioned close to the joint area. The same is true for massive nylon and acetal parts. Positioning the joint 1/4 in. or less from the area of horn contact is termed "near-field welding" and more than 1/4 in. from the contact area is called "far-field welding." In general, the closer the horn can be applied to the joint area, the faster an acceptable weld can be accomplished.

Materials rated highly for ultrasonic welding include polystyrene, SAN and polycarbonate. Materials rated excellent, but where part structure is important, include rubber-modified polystyrene, ABS, and acrylics. Resins that perform highly in near-field welding, but which may be limited in far-field applications,

Table 8.2
Tips for Joining of Plastics

- Avoid altering the horn in any way—doing so may damage it or change its resonating characteristics.
- Use caution in specifying materials to be welded—only like materials or certain similar plastics with common melting temperatures and polymer construction such as ABS, SAN, styrene, and acrylic can be welded.
- Design parts specifically for ultrasonic assembly.
- Avoid silicone mold releases.
- Fixture parts properly for optimum results.
- Test materials regularly to be sure of consistency and melt temperature.
- Avoid *excessive* regrinds, which may alter strength characteristics. However, to date, regrinding has not been an extreme problem. A pragmatic approach is suggested for the uninitiated. Experience will usually obviate the trial-and-error process.
- Determine exact weld time, hold time, and pressure by trial and error.

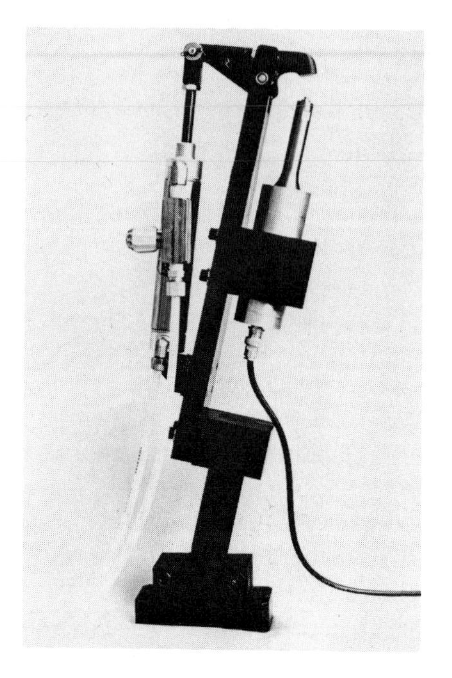

Figure 8.4. Each welding head can be adjusted in three planes to follow the contours of the part.

are the nylons, olefins, and acetals. Materials that must be selected critically for ultrasonic welding are butyrates, cellulosics, acetates, and vinyls. Some formulations perform well while others may degrade either physically or chemically. Films and additives modify basic properties of materials. Glass fillers usually enhance the weldability, yet too much may be detrimental. Acrylic is now considered a good welder both near- and far-field. Fillers, not films, modify basic properties of materials.

Generally, dissimilar plastic materials cannot be welded because with different melt temperatures, bonding will not take place. In addition to melt temperature, similarity in chemical structure is required. Plastic materials with similar melt temperature and like chemical structure that can be welded interchangeably include ABS, SAN, styrene, and acrylic.

Plastics to be welded ultrasonically should have their joints designed before the mold is cut. Revising existing molds is usually possible, but effectiveness may suffer due to necessary compromise. In ultrasonic welding of plastics, a small triangular projection molded into a component, called the energy director, focuses energy to a concentrated area to produce an almost instantaneous melt.

The melt flows evenly throughout the joint interface, bonding the two surfaces.

The energy-director volume should provide sufficient molten material to coat the joint area adequately. The energy director height should be at least 0.005 inches, or one-tenth the joint width. The energy director width should be at least one-fifth the joint width. Two or more energy directors should be used with thick walled joints when the energy director height of one-tenth the joint width is greater than 0.020 inches, and the sum of their two heights should equal one-tenth the joint width.

Types of joints that may be welded ultrasonically include butt, step, and tongue-and-groove. Joint strength appearance requirements, and machining and molding capabilities, determine the type of joint. Tongue-and-groove joints have good strength characteristics, but the need to maintain clearance on the

Figure 8.5. Continuous seam welding. A white thermoplastic kitchen door is about to pass under compression roller and stationary welding horns.

joint sides makes it more difficult to mold. The step joint molds readily, and provides a strong, well-aligned joint with minimum effort. Joint surfaces should not fit so tightly that relative vibratory motion is restricted.

Welding Applications. Applications of ultrasonic welding of plastics are numerous and can be found in products of all sizes and descriptions—wherever two or more plastic components are specified. Following are examples of a number of welding applications:

- Contoured impact styrene parts for a toy camera are aligned and joined in a contoured nest.
- Side rails are joined to a light diffuser. The crystal polystyrene parts are clamped in a special fixture that indexes the part so it can be welded in several places.
- Polycarbonate clear lens and water meter case are hermetically sealed. The ultrasonic energy is set to turn on as the horn contacts the part.
- ABS food warmer cover and plate are hermetically sealed to withstand automatic dishwashing.
- Polystyrene upper and lower parts of a baby food dish are hermetically sealed.
- Woven polypropylene fabric is seamed and slit continuously to eliminate end unraveling. Fabric is fed under the sealing surface of the horn maintaining a set gap between horn and anvil. The anvil is angled so that after the seal takes place, the cutting surface will part the seal cleanly. Feed rates of 50 fpm are achieved.
- Flashcube styrene base with bulbs is welded to a clear styrene outer case.
- Polysulfone two-channel splice case for aircraft connections is welded in a tongue-and-groove joint.
- PVC sleeve is welded over a terminal-ended wire with PVC insulation.
- Impact polystyrene flashlight case sections are joined.
- Polystyrene film frames are welded to contain film. Two slides are welded simultaneously.
- A film pouch of 0.003-in. thick acetal is sealed at the rate of 50 fpm.
- Glass-filled SAN discharge valve for an automatic washing machine is welded.
- ABS cover and body for an auto vacuum manifold assembly are joined.

Figure 8.6. Large, flat parts are conveyor-fed under the stationary horns, which provide continuous ultrasonics. Feed rates of up to 15 ft./min. are possible for applications such as furniture and door panels.

- Plastic sheets resembling fabrics are "stitched" together in a wide variety of patterns with a sprocket-horn system.
- Components for an ABS handle are welded. The 7½-in. long prism shaped sections are joined with two welds at each end. A separate horn "hilt" is made at each end.
- Ends of a polyester sheet are seamwelded to form a belt.
- A polyethylene oxide transformer case and back are welded in four places simultaneously.

Inserting Metal Parts into Plastics

In ultrasonic insertion of metal pieces in plastics, the horn also vibrates at 20 hKz and drives the insert into a pilot hole slightly smaller in diameter. Movement created between the insert and the plastic melts an annular ring of plastic allowing the insert to slip in. The displaced melted plastic flows into properly designed undercuts, flutes or knurls, hardens, and gives a good mechanical lock.

Economy is said to be the most important advantage of ultrasonic insertion. Molding-machine operators are not required to place inserts into molds, permitting high-speed automatic molding operations. Mold cycle time is reduced along with the possibility of mold damage. Also, freeing the mold designer of molded-in insertion requirements increases latitude in design. Costs can be reduced further with ultrasonic inserting since tolerances are not critical. Standard screw-machine part tolerances are adequate.

Another advantage to ultrasonic inserting is that very little stress exists around the insert. During inserting, parts remain relatively cool, since heat is generated only at the plastic/metal interface. Still another advantage is that ultrasonic inserting can be accomplished by contacting either the metal or the plastic. In some applications, it may be desirable to contact the plastic and melt the plastic around the insert.

Many different plastics are highly suitable for ultrasonic inserting. Plastics rated excellent for this use are unfilled rubber-modified and glass-filled polystyrene, SAN, ABS, polycarbonate, nylon, polysulfone, acetal, acrylic, polypropylene and polyethylene.

Design considerations vary considerably in ultrasonic inserting. The type of metal and plastic used, the size and shape of the insert, the application and

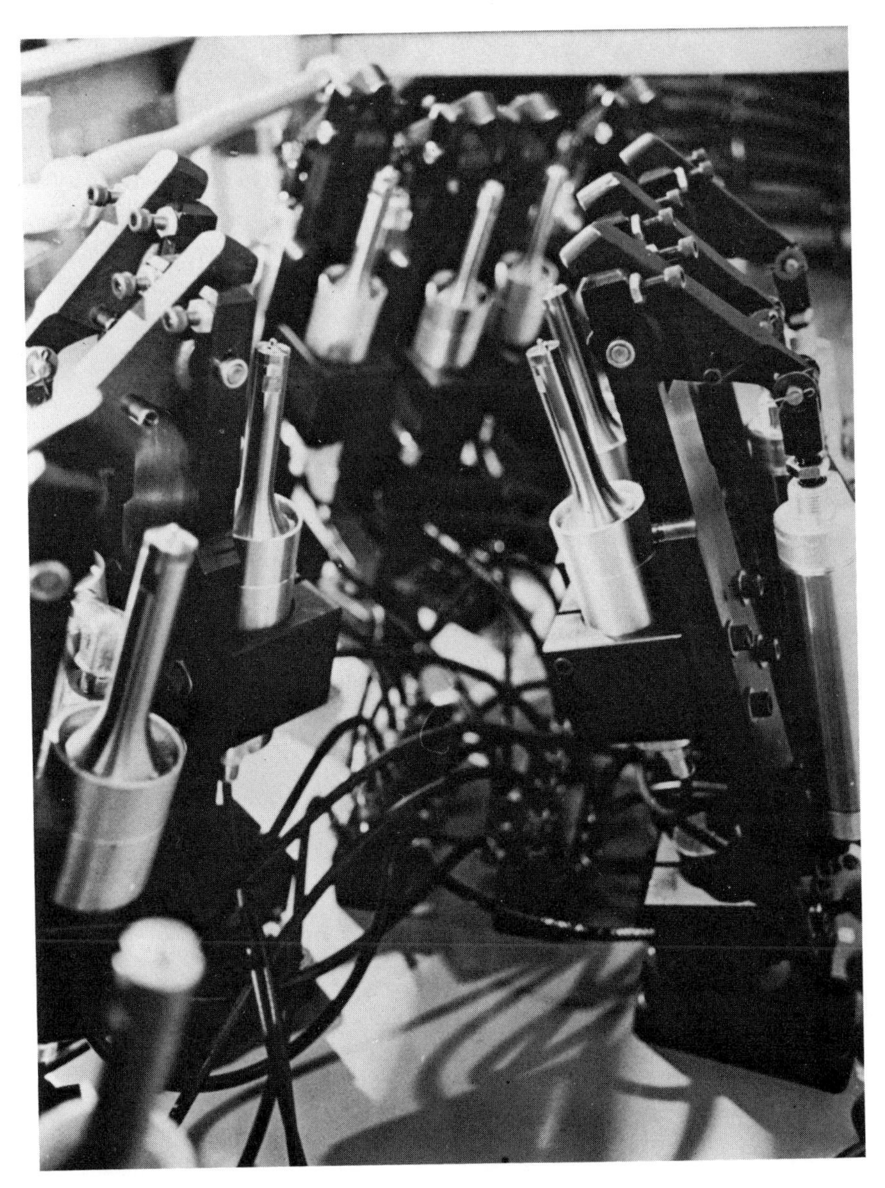

Figure 8.7. This gang-welding system consists of eleven welding heads that are simultaneously activated by two TW-8 power supplies to ultrasonically assemble heater ducts.

strength required, along with other factors make it unwise to set forth absolute standards for any application. However, several studies have determined optimum conditions for specific applications. One study found that when inserting internally threaded inserts into polycarbonate bosses, the boss diameter should be a minimum of twice the insert diameter. Another study showed that with a certain steel insert, optimum conditions were reached when the insert melt zone diameter was 0.045 in. less than the hole diameter, and the insert's bushing diameter was 0.015 in. greater than the hole diameter. The insert size and the number of inserts to be driven simultaneously determine the ultrasonic exposure time.

Insertion Applications. Inserts that may be driven ultrasonically include a wide variety of bushings, hubs, ferrules, terminals, feed-through fittings, pivots, retainers, fasteners, hinge plates, handles, locating pins, binding posts, and ornamental or decorative attachments. Following are some examples of insert applications:

- A hinge is inserted into each end of an acetate eyeglass frame using a horn with a hole drilled through it for attachment to a vacuum line to facilitate placement of the small hinges. A removable horn tip has a cavity that fits around the hinge. The hinge is placed into the tip by hand and the vacuum holds it in place until it is inserted into the frame.
- A lipstick metal tube is inserted into a styrene base with a special alignment stud to hold the part during insertion. The horn contacts the base while the tube is held on a mandrel.

Staking Employs Same Principles

As with ultrasonic welding and inserting, ultrasonic staking of plastics employs the same principle of creating localized heat through the application of high-frequency vibrations. Staking utilizes the reshaping of plastic projections to fasten the plastic to another object, usually metal.

In staking, a stud is reshaped into a locking head to retain another material. A specially contoured horn contacts and melts the stud, forming the locking head.

The advantages of staking are numerous. The process is extremely rapid, since only minute portions of plastic are reshaped. Consistent results are achieved and tight assemblies are assured because there is no recovery due to material memory. Staking is clean since no plastic adheres to the horn, which remains relatively cool. There is no apparent degradation of the plastic material, since it is made fluid enough to flow just before it reaches melting temperature. This is particularly important with nylon, which degrades quickly if it is heated beyond its melting temperature.

In general, plastics ideally suited for ultrasonic staking and swaging are identical to those suited for inserting.

Staking Applications. Typical applications for ultrasonic staking include automobile instrument clusters and taillight assemblies, a broad range of components for the appliance industry, including radio and television panels, and electronic components. Following are some application examples of staking:

- A small electrical switch blade is staked to a nylon insulation spacer.
- A leaf-type contact assembly is staked into a filled nylon timed extension cord base.
- A metal back is staked onto a timer housing in three places with breakaway studs. Each stud has a hole molded into it that is approximately one-half the stud diameter. Should the timer require repair, the studs are broken off to remove the plate. The plate is reassembled with self-tapping screws.
- A metal mainspring watch tension control is staked onto a tiny 40% glass-filled nylon boss at a frequency of 40 kHz using a sonic converter-horn unit. Stud dimensions are approximately 0.050 in. by 0.008 in. A microscope is used to fixture the assembly properly. The head is cooled under pressure after the vibrations stop, to assure a tight assembly.
- Two stakes hold a polypropylene slide to the contact of a flashlight switch.

Spot Welding

Vibrating ultrasonically, the pilot of the tip passes through the top sheet and enters the bottom sheet to a depth of one-half the top sheet thickness. The molten plastic displaced is shaped by a radial cavity in the top and forms a neat raised ring on the surface. Simultaneously, energy is released at the interface, producing essential frictional heat. Molten plastic displaced from the second sheet flows into the preheated area and forms a permanent molecular bond.

Table 8.3
Plastics Weldability Chart

Material	% Weld* Strength	Spot Weld	Staking & Inserting	Swaging	Welding** Near-Field	Welding** Far-Field	Remarks
Polystyrene unfilled	95–100+	E	E	F	E	E	Excellent acoustical properties; produces strong smooth joints.
Structural foam	90–100	E	E	R	G	P	High-density foams weld best.
Rubber modified	95–100	E	E	G	E	G–P	Welding characteristics depend on degree of impact resistance.
Glass-filled	95–100+	E	E	F	E	E	Weldable with filler content up to 30%.
SAN	95–100+	E	E	F	E	E	Particularly good as glass-filled compounds.
ABS	95–100+	E	E	G	E	G	Can be bonded to other polymers such as SAN, styrene, acrylics.
ASA	95–100+	E	E	G	E	G	Can be bonded to other polymers such as SAN, styrene, acrylics. Weld characteristics depend on degree of impact resistance.
XT-Polymer	95–100	E	E	G	E	G	Can be bonded to other polymers such as SAN, styrene, acrylics.
Polycarbonate	95–100+	E	E	G–F	E	E	High melting temperature requires high energy levels. Oven-dried or "as-molded" parts perform best due to hygroscopic nature of the material. Moisture will inhibit welds.
Nylon	90–100	E	E	F–P	G	F	
Vykan-A	90–100	E	E	F	P	G–F	
Polysulfone	95–100+	E	E	F	G	G–F	
Cycoloy 800	95–100+	E	E	G	E	G	

Acetal	65–70	G	E	P	G	G	Requires high energy and long ultrasonic exposure because of low coefficient of friction. Watery at melt.
Polyimide	80–90	F	G	P	G	F	High temperature polymer only, injection grade weldable. Too brittle for staking.
Acrylics	95–100+	G	E	P	E	G	Weldable to ABS and SAN. Cast grades more difficult to weld due to high molecular weight.
Cycovin	95–100+	E	E	G	G	F	Alloyed polymers require high degree of energy control or decomposition can take place within joint, resulting in weak or poor welds.
Kydex	95–100+	E	E	G	G	F	
Polyphenylene oxide	95–100+	E	G	F–P	G	G–F	High melting temperatures require high energy levels.
Noryl	95–100+	E	E	F–P	G	E–G	
Phenoxy	90–100	G	E	G	G	G–F	
Polypropylene	90–100	E	E	G	G–P	F–P	Horn design for welding is particularly critical; filled compounds usually better, but need individual testing.
Polyethylene	90–100	E	E	G	G–P	F–P	
Structural foam	85–100	E	E	F	G	F–P	
Butyrates	90–100	G	G–F	G	P	P	Weldability varies with formulation and part configuration; however, these materials usually perform well in staking, spot welding, and inserting applications. Decomposition of some formulations may occur.
Cellulosics	90–100	G	G–F	G	P	P	
Vinyls	90–100	G	G–F	G	F–P	F–P	
Glass-filled acetal	70–75	G	G–F	P	G	G	Watery at melt. Small assemblies best.

(*continued*)

Table 8.3 (continued)
Plastics Weldability Chart

Material	*% Weld* Strength*	*Spot Weld*	*Staking & Inserting*	*Swaging*	*Welding ***		*Remarks*
					Near-Field	*Far-Field*	
Glass-filled ABS	95–100	E	E	G	E	G	Weldable to acrylics and polystyrene, SAN and ASA. Extruded sheet more difficult to weld due to lack of energy directors. Up to 30% glass. Over 20% glass will cause wear.
Rubber-modified acrylic	95–100	E	E	G	E	G	Weldable to ABS, SAN, and ASA. Cast sheet more difficult to weld due to lack of energy directors; also, usually has high molecular weight causing it to be more brittle in thick sections.
Cellulose acetate Cellulose acetate butyrate Cellulose propionate	90–100	G	E	G	G	F	Energy levels must be controlled, otherwise degradation may occur.
Chlorinated polyether Ethyl cellulose Ethyl vinyl acetate		Considered Poor Welders					Corrosion-resistant. Similar to polyethylene. Near-field weld applications. Poor acoustic properties.
Fluorocarbons		Not Usually Weldable					Due to low coefficient of friction, not weldable. Poor acoustical properties. Fillers increase weldability.
Glass-filled polyamide	90–100	E	E	G–F	E	F	Hygroscopic, i.e. ABS or BS moisture, which will affect welds.

Polyallomer	95–100	E	E	G–F	E	F	Properties similar to polyethylene. Near-field weld applications.
Glass-filled polycarbonate							Hygroscopic, i.e. absorbs moisture, which will affect welds. Dry before welding. Up to 30% glass.
Glass-filled polyethylene	90–100	E	E	G	G–P	F–P	Due to poor acoustic properties, successful only with near-field welds. Fillers improve weldability. Up to 30% glass.
Glass-filled polypropylene	90–100	E	E	G	G–P	F–P	Poor acoustical properties. Near-field welds only. Fillers improve weldability. Up to 30% glass.
Methyl pentene polymer	90–100	E	E	G	G–P	F–P	Transparent, heat-resistant polyolefin.
Glass-filled polystyrene	95–100+	E	E	F	E	E	Excellent acoustical properties. Weldable to SAN, ABS, and ASA. Up to 30% glass.
Polyurethane	Poor Welder, Usually Considered Near Thermoset, Sheet or Films Weld O.K.						Softening point is not far below its decomposition point.
PVC	90–100	G	G–F	G	F–P	F–P	Softening point is not far below its decomposition point. Plasticizers inhibit weldability. Also may generate acids. Welds deteriorate in storage due to migration of plasticizers back to joint.
Glass-filled SAN	95–100+	E	E	F	E	E	Weldable to acrylics, styrene and ASA. Up to 30% glass.
Ionomer	90–100	E	E	F–P	G	F	Only near-field welding recommended.

E = Excellent; G = Good; F = Fair; P = Poor.

*Weld strengths are based on destructive testing. 100+% results indicate that parent material of plastic part gave way while weld remained intact.

**Near-field welding refers to joints 1/4-in. or less from area of horn contact; far-field welding to joint more than 1/4-in. from contact

Table 8.4
Ultrasonic Welding Compatibility of Thermoplastics (X denotes compatibility; O denotes some, but not all, grades and compositions compatible).

	ABS	*ASA*	*XT-Polymers*	*Acrylics*	*Cycoloy-800*	*Polycarbonate*	*Nylon*	*Vykan-A*	*Acetal*	*Polyethylene*	*Polypropylene*	*Polystyrene*	*Polysulfone*	*SAN-NAS*	*PVC*	*PPO*	*Noryl*	*Cycovin*	*Kydex*	*Polyimide*	*Butyrate*	*Cellulostics*
ABS	X	O	O	X	X							O		O				O	O		O	
ASA	O	X	O	O	O							O		O				O	O			
XT-Polymer	O	O	X	O	O							O		O				O	O			
Acrylics	X	O	O	X	O	O								O				O	O			
Cycoloy-800	X	O	O	O	X	X								O				O	O			
Polycarbonate				O	X	X																
Nylon							X	O														
Vykan-A							O	X														
Acetal									X													
Polyethylene										X												
Polypropylene											X											
Polystyrene	O	O	O									X		O			X					
Polysulfone													X									
SAN-NAS	O	O	O	O	O							O		X			O					
PVC															X			O	O			
PPO																X	X					
Noryl												X		O		X	X					
Cycovin	O	O	O	O	O													X	O			
Kydex	O	O	O	O	O										O			O	X			
Polyimide																				X		
Butyrate	O																				X	
Cellulosics																						X

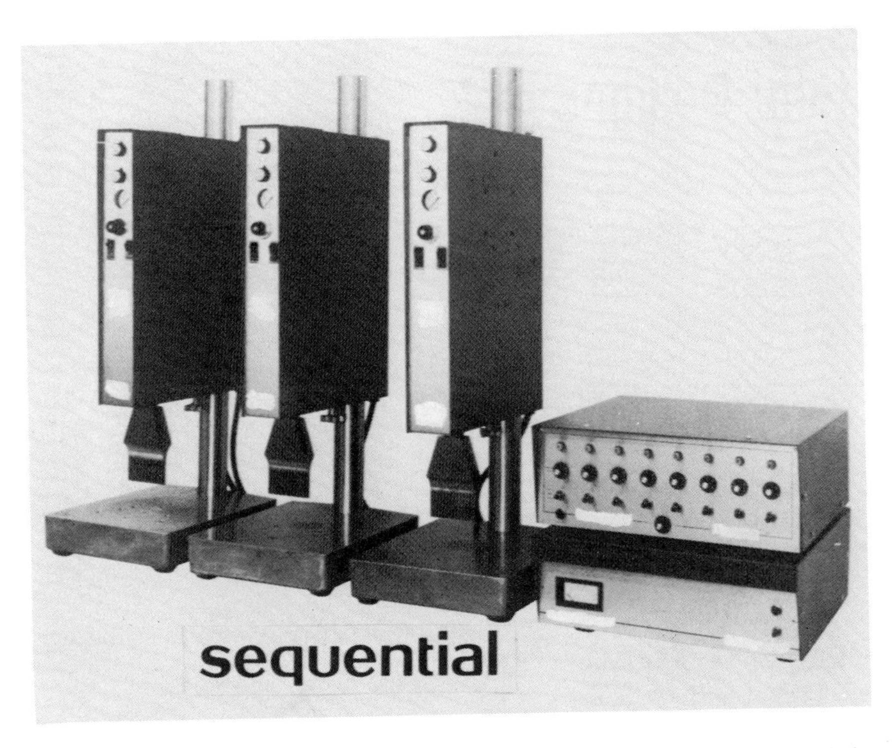

Figure 8.8. Typical layout of a sequential three-station welding system, tooled for three independent welds.

Materials may be ABS, PVC, polypropylene, or almost any other thermoplastic. Both fixed and portable ultrasonic assembly systems may be used for spot welding. Portable units are ideal for large parts too bulky for fixed units, and parts with hard-to-reach joining surfaces.

Spot Welding Applications. Spot welding may be used to assemble curtain wall panels, appliances, aircraft ductwork, furniture, car bodies, snowmobiles, amphibious vehicles, siding for houses, gutters, downspouts, animal shelters, toolsheds, trailer bodies, storage facilities, and for any other applications that require the assembly of large thermoplastic parts.

Chapter 9
Joint Designs

Although ultrasonic welding is a relatively new technique, much has been written about its advantages as a reliable and economical method for joining engineering plastics. It is, in fact, ideally suited for use with acetal, nylon and acrylic resins. But careful joint design is highly important to the success of this joining technique.

The Butt and Peak Joint

Until recently, in ultrasonic welding the most commonly used joint design was the butt-and-peak joint, or an especially strong variation, the tongue-and-groove joint, as shown in Figure 9.1.

Although the butt joint is simple to design, it is not preferred for use with acetal and nylon resins because of the V-shaped bead (energy director) which flows away from the weld and crystallizes before sufficient heat is generated to

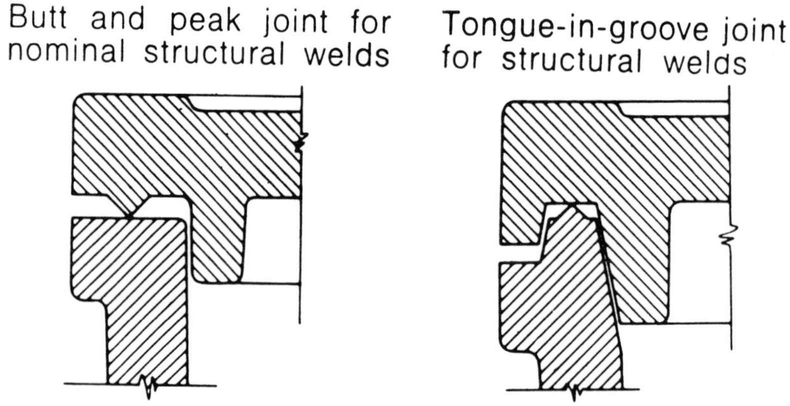

Figure 9.1. Basic joint designs used for welding pastics.

weld the full width of the joint. Consequently, with the exception of the area immediately adjacent to the energy director, only spotty welding occurs. Because of the open joint, the melt is exposed to air during welding which accelerates the crystallization and, with nylon, can cause degradation by oxidation.

For these reasons, the butt joint is best used for parts designed with interrupted joints requiring only nominal strength, or on large parts where part-to-part dimensional variations or joint design limitations are encountered.

The Shear Joint

While the butt-and-peak joint works in many applications, its drawbacks, as mentioned above, are numerous. Engineers at one manufacturer's European affiliate were instrumental in remedying this problem when they introduced the shear joint in 1967. An ever-growing list of applications for the joint has followed and many more are underway. Typical are the applications shown in Figures 9.2 and 9.3.

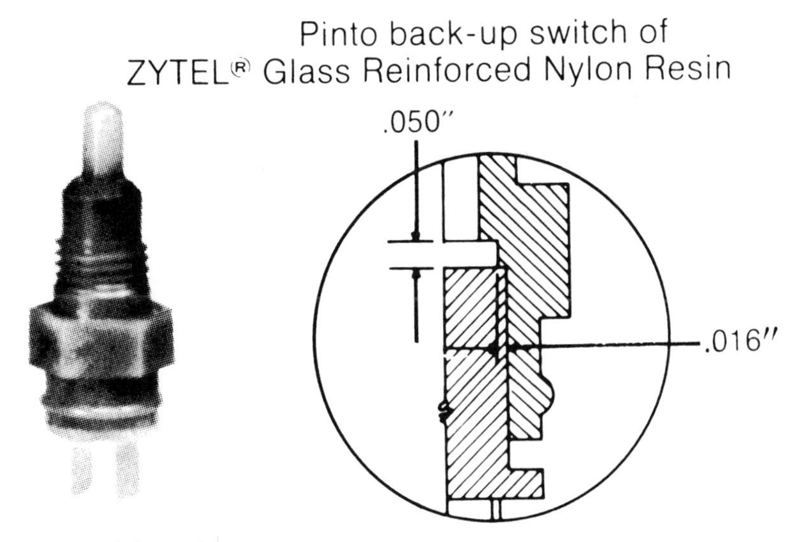

Figure 9.2. Shear joints are used for positive, permanent welding of plastics components.

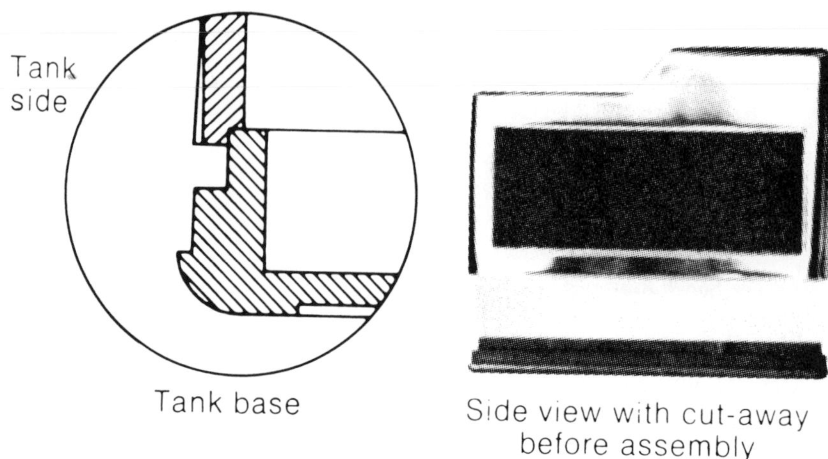

Figure 9.3. Plastics surfaces continue to melt as two surfaces of shear joint slide together.

The basic shear joint is shown in Figure 9.4, before, during and after welding. It is the preferred joint for the ultrasonic welding of acetal, nylon and acrylic resins. Figure 9.5 shows three variations of this basic joint.

In forming the joint, initial contact is limited to a small area, which is usually a recess or step in either of the parts. Welding is accomplished by first melting the contacting surfaces. Then, as the parts telescope together, they continue to melt along the vertical walls. The smearing action of the two melt surfaces eliminates leak and voids, making this the best joint for leak-free seals.

The shear joint has the lowest energy requirement and the shortest welding time of all joints. The low initial energy requirement is due to the small contact area, which remains relatively constant as the parts telescope together. Heat generated at the joint is retained until the vibrations cease, because during the telescoping and smearing action, the melted plastic is not exposed to air, which would cool it too rapidly.

Weld strength is determined by the depth of the telescoped section, which is a function of the weld time and part design. Joints can be made stronger than the adjacent walls by designing the depth of telescoping as 1.25 to 1.5 times

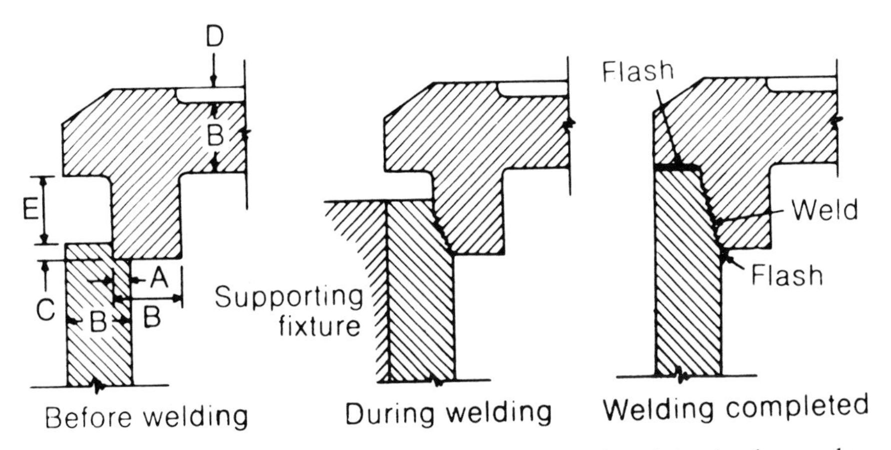

Figure 9.4. The surface area of contact increases as a shear joint develops under ultrasonic energy.

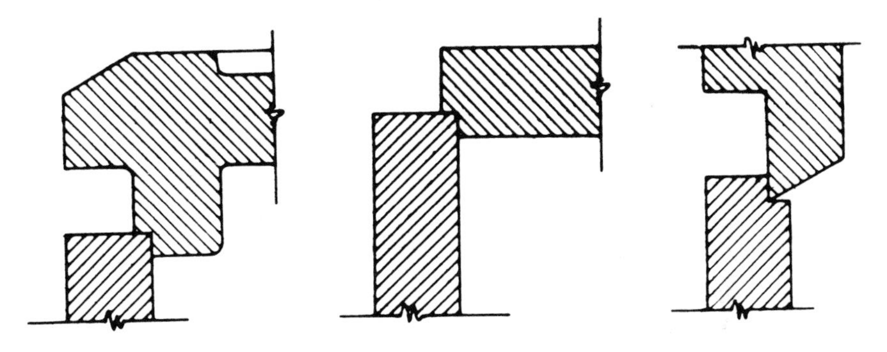

Figure 9.5. Three variations of the basic shear joint for plastics.

the wall thickness. The tensile strength of the welded joint is seldom equal to a continuous strip of material.

Several important design aspects of this joint must be considered. The top part should be as shallow as possible, in effect just a lid. The walls of the bottom section must be supported at the joint by a holding fixture in order to avoid expansion during welding. Non-continuous or inferior welds will result if the upper part slips off the lower part, or if the stepped contact area is too small. Modifications to the joint, shown in Figure 9.6, should be considered for large parts because of dimensional variation, or for parts where the top piece is deep and flexible.

Allowance must be made in the design of the joint for the flow of molten material displaced during welding. When flash cannot be tolerated, a trap similar to the one shown in Figure 9.7 can be designed into the joint.

Basic Design Factors

The diagrams (Figure 9.8) show time-temperature curves for a common butt joint and the more ideal joint incorporating an energy director. This modified

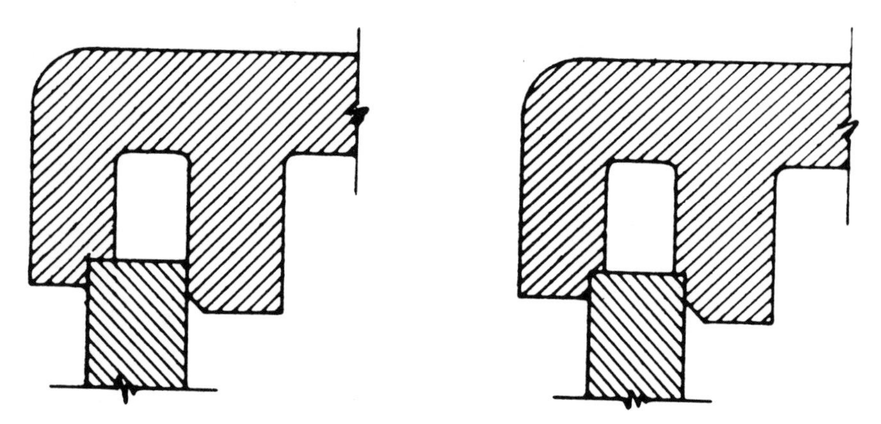

Figure 9.6. Modifications of the basic shear-joint designs should be considered for large parts, due to dimensional variations.

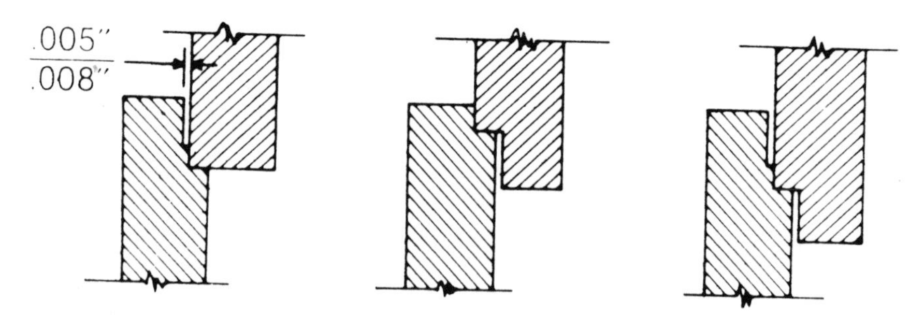

Figure 9.7. Flash traps can be designed into shear joints when flash cannot be tolerated.

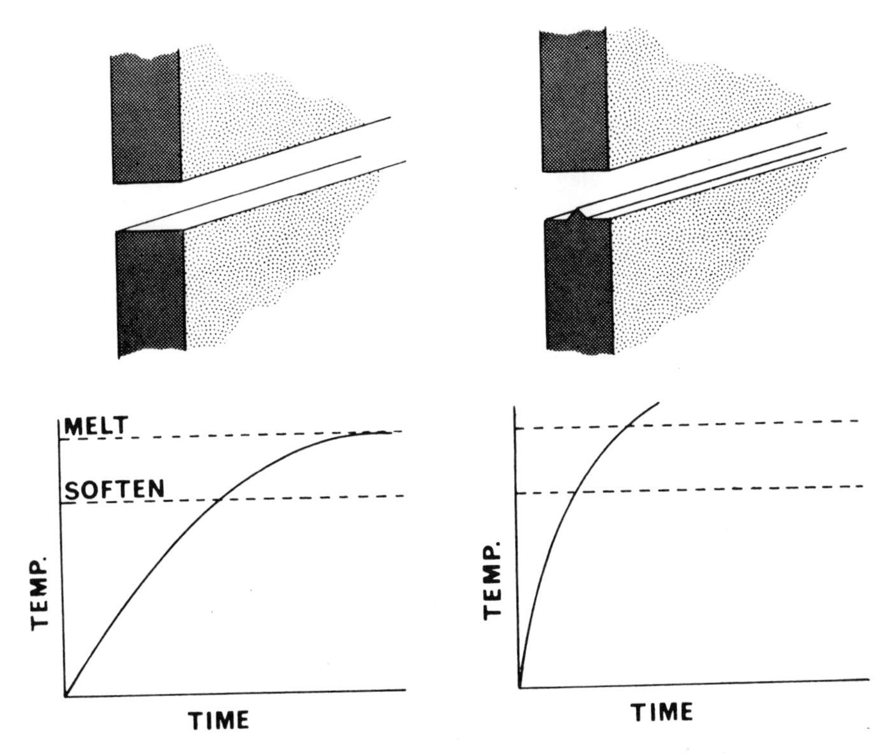

Figure 9.8. Temperature of the joint line develops more quickly when an energy director is designed into the joint.

joint permits rapid welding while achieving maximum strength. The material within the director becomes the sealant, which is spread throughout the joint area as indicated below.

Figure 9.9 illustrates a simple butt joint modified with an energy director, showing desired proportions before weld and indicating the resultant flow of material. Parts should be dimensioned to allow for the dissipation of the material from the energy director throughout the joint area as illustrated.

Figure 9.10 illustrates a step joint used where a weld bead on the side would be objectionable. This joint is usually much stronger than a butt joint, since material flows into the clearance necessary for a slip-fit, establishing a seal that provides strength in shear as well as in tension.

A tongue-and-groove joint (Figure 9.11) usually has the capability of providing greatest strength. The need to maintain clearance on both sides of the

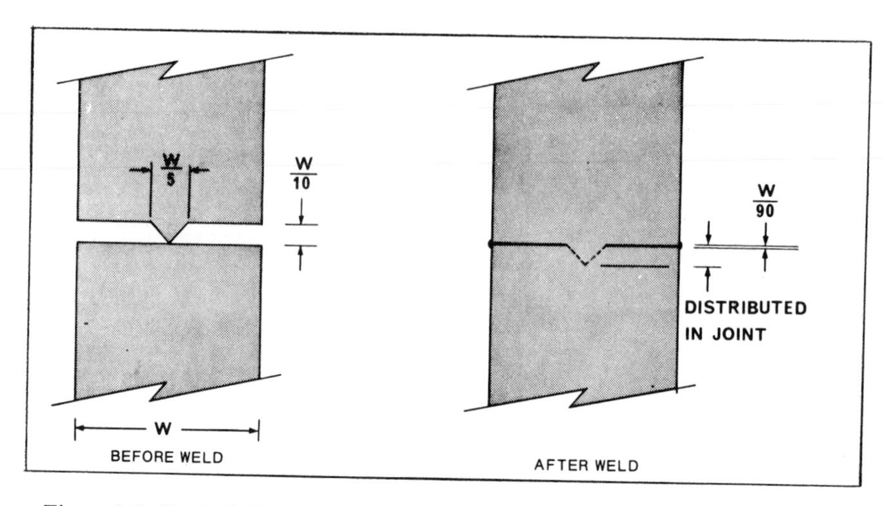

Figure 9.9. Typical dimensions of an energy director before and after welding.

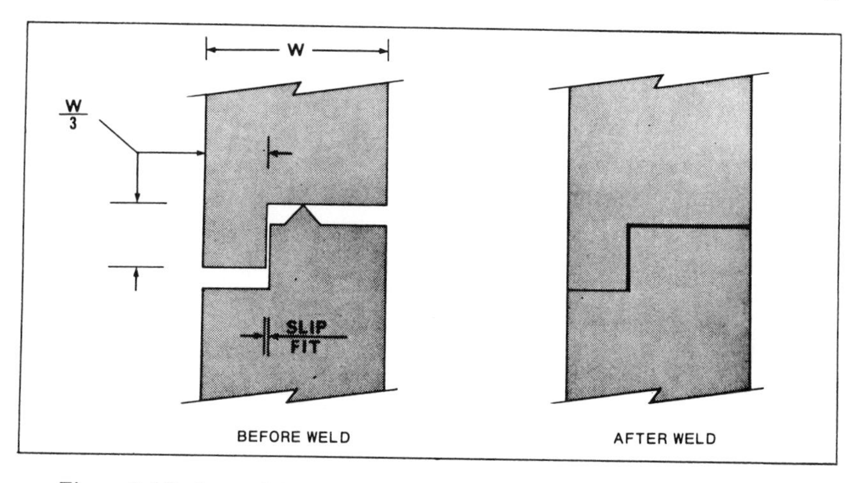

Figure 9.10. A step joint is used where a weld bead on the side would be objectionable.

tongue, however, makes this more difficult to mold. Draft angles can be modified concurrent with good molding practices, but interference between elements must be avoided.

Figure 9.12 illustrates basic joint variations suitable for ultrasonic welding. These are suggested guidelines for typical joint proportions. Specific applications

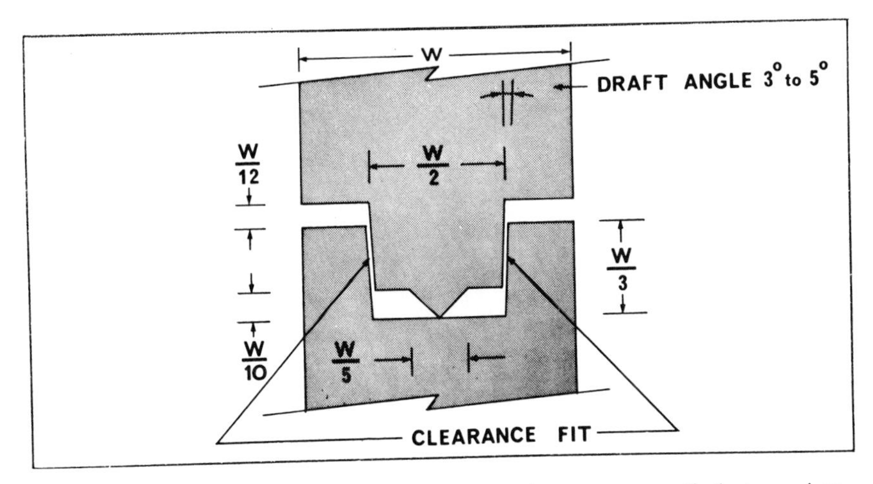

Figure 9.11. A tongue-and-groove joint provides great strength, but requires accurate molding.

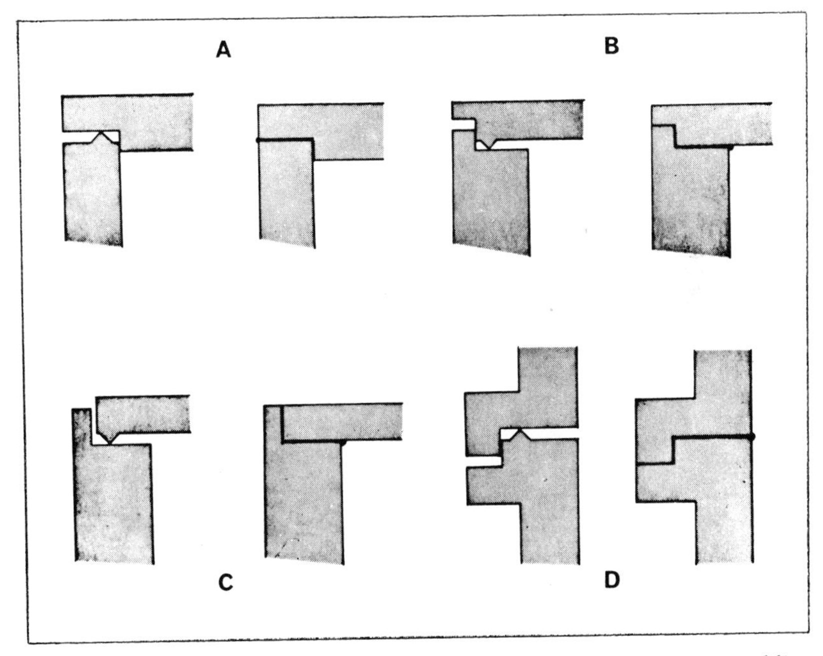

Figure 9.12. A great variety of joint designs is suitable for ultrasonic welding, depending on the specific application requirements.

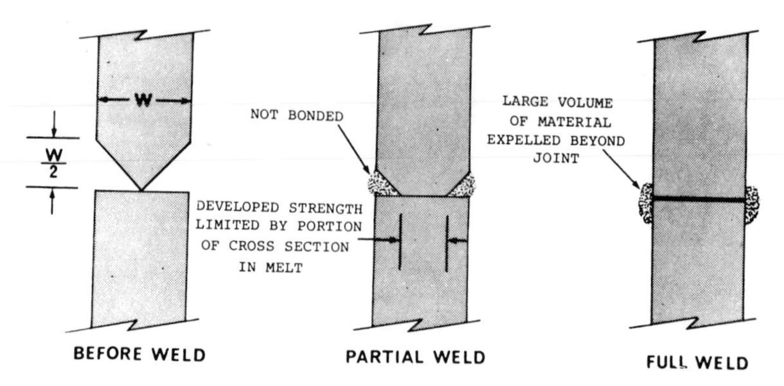

Figure 9.13. Beveling one joint face should be avoided, because it results in either a low-strength joint or excess flash.

may require slight modification. Practical considerations suggest a minimum height of 0.005 in. for the energy director. Where height greater than 0.020 in. is indicated, two or more directors should be provided, with the sum of heights equalling the formula dimension.

Factors to be Avoided

A typical mistake is beveling one joint face at 45-degree angles. Figure 9.13 shows the result if this practice is followed. It should also be noted that joints designed for solvent sealing can generally be modified to meet ultrasonic welding requirements.

Chapter 10

Rotary Processing Techniques

Introduction to the Rotary Ultrasonic Machine

The rotary ultrasonic machine utilizes the advantages of both high-speed rotary cutting tools and reciprocating ultrasonic tools. In operation, a diamond abrasive cutting tool is simultaneously rotated and ultrasonically vibrated. This technique provides accelerated cutting and a greater degree of precision than is possible with conventional tools for drilling, threading, grinding and milling of hard, brittle materials such as aluminas, glass, ferrite, quartz, zirconium, ruby, sapphire, beryllium oxide, boron composites, various thermoset and thermoplastic resins, and hard rubber-resin combinations. Since no abrasive slurry is used, deep holes can be drilled to close tolerances at relatively high speeds without taper.

How the Rotary Ultrasonic Machine Works

An automatic-tuning power supply provides 300 watts of electrical energy to a sonic converter which converts this energy into 20 kHz mechanical vibrations. These vibrations are amplified and applied to the rotating cutting tool by means of a special alloy horn or mechanical impedance transformer.

With the diamond abrasive tool simultaneously rotating at speeds up to 5,000 rpm (variable) and vibrating, the friction between the tool and the workpiece is minimized. This accelerates the cutting and reduces the required pressure. The application of ultrasonics to the rotating diamond tool greatly assists the machining operation.

The resultant reduction in friction between the tool and workpiece provides faster and smoother cutting. It eliminates binding and loading of the tool and enables cutting at lighter tool pressures. This combination of reduced friction and lower tool pressure not only extends tool life but also permits the machining

Figure 10.1. Rotary welding of a bottom section in plastics materials.

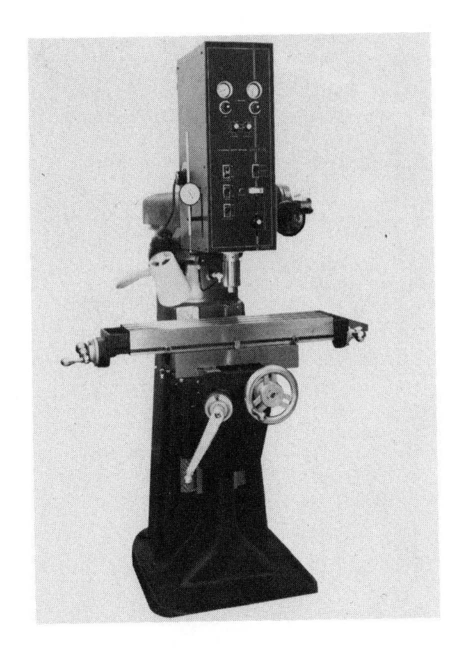

Figure 10.2. Vertical rotary welding machine set up for a reaming application.

of delicate components without cracking, and minimum shelling at the point of entry and breakthrough.

Equipment Features

- Head unit tilts 90 degrees to either side of center.
- Automatic or manual head downfeed system can be cycled for fully automatic operation.
- Motor speed control 0–5,000 rpm.
- 300 watt variable power supply.

Applications of a Rotary Ultrasonic Machine

High alumina ceramics, technical ceramics, ferrites, porcelain, glass, boron tungsten laminates and other hard, brittle materials that are usually difficult to machine can be drilled, ground (internal and external), end milled, threaded (internal and external) and trepanned with a rotary ultrasonic machine. The abil-

Figure 10.3. Threading a ceramic component after firming.

ity to machine these materials provides a unique facility for the production of components with, as in the case of ceramics, greater precision than could be guaranteed with a part machined before firing. It is ideal for prototype work and enables the user to modify preformed parts. These are typical of the many applications for which a rotary ultrasonic machine offers advantages.

Advantages of a Rotary Ultrasonic Machine

- Accelerated cutting, i.e. ⅛-in. diameter hole through ¼-in. thick, 99.9% aluminas in 20 seconds.
- No tool loading.

Figure 10.4. Typical rotary automatic table-feed unit for a single welding station.

Table 10.1
Typical Cutting Rates–Drilling.

Material	*Diameter of Drill*	*Depth of Cut*	*Approximate Time Required*
99.9 Alumina	0.042	0.250	1 min., 50 sec.
	.125	.250	14 sec.
	.250	.250	11 sec.
	.375	.250	16 sec.
	.500	.250	18 sec.
Glass	.042	.500	1 min., 15 sec.
	.250	5.000	2 min., 10 sec.
Ferrite	.080	.250	23 sec.
	.125	.250	19 sec.
	.250	.250	13 sec.
Boron Composite	.500	.700	48 sec.
	.250	.500	19 sec.
	.125	.500	26 sec.

- Capable of drilling deep and small holes without taper.
- No core seizure when core drilling.
- Tool will cut at low pressure eliminating tool wander when drilling small holes.
- Longer tool life.
- No abrasive slurry required.
- No binding, seizing or loading of the tool on deep-hole drilling.

Chapter 11

Ultrasonic Degating of Molded Parts

Ultrasonic welding machines, which normally are used to join plastics parts, can also be used to separate injection molded parts from their runner system. The ultrasonic degating results in a relatively smooth surface finish at point of separation (Figure 11.1) making deburring or other finishing operations unnecessary. For low-quality moldings, where appearance is unimportant, ultrasonic degating may be less attractive for economic reasons.

Development of the process grew from observations that molded parts sometimes will separate at thin sections during ultrasonic welding. Thin gate sections apparently function in a manner similar to energy directors in ultrasonic welding. Energy directors are thin, pointed ridges molded into parts along edges where ultrasonic welding is to occur. Ultrasonic energy becomes concentrated at these ridges, causing them to melt: thin gate sections melt in the same manner.

Rigid Plastics Work Best

Plastics that are ultrasonically weldable are usually also ultrasonically degateable. Included are PS, ABS, PC, SAN, PPO, nylon, acetal, polysulfone, polyimide, and acrylic.

Very brittle or very soft plastics do not give good results. Brittle thermoplastics tend to fracture mechanically before melting; thermosets, of course, do not melt at all, but a mechanical break is still possible. Soft and very flexible plastics, such as LDPE, do not transmit ultrasonic energy efficiently. This same limitation prevents degating while the runner system is still hot after ejection

This chapter is exerpted from an article by J. R. Sherry of Bronson Sonic Power Company that originally appeared in the April 1970 issue of *Modern Plastics*. Reprinted by permission of Modern Plastics, McGraw-Hill, Inc.

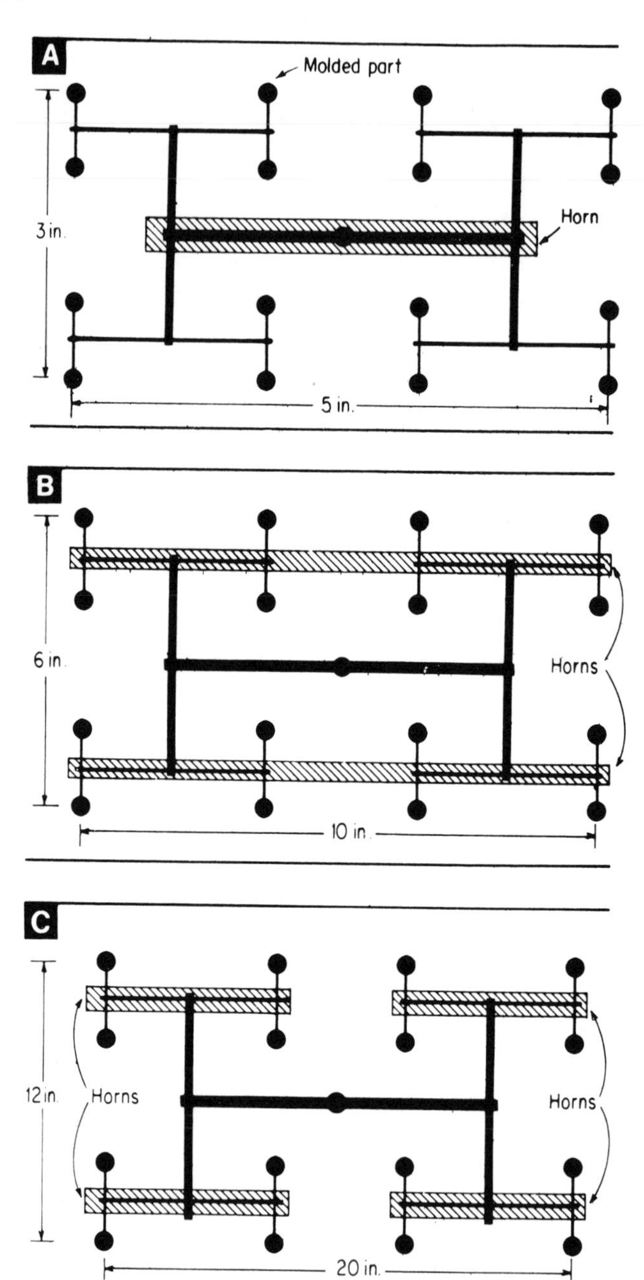
A
Molded part
Horn
3 in.
5 in.
B
6 in.
Horns
10 in.
C
12 in.
Horns
Horns
20 in.

from the mold. Addition of fillers and reinforcements improves the processing of flexible materials by stiffening them.

Operating Procedure Similar to Welding

Anyone who has an ultrasonic welder can also do ultrasonic degating with very little added equipment. An ultrasonic welder with a power rating of 2200 in.-lb./sec. (250 w.) is adequate for most jobs. Higher-powered welders normally can handle heavier gate cross-sections.

A runner system as it comes out of the mold (i.e. with parts attached to it) is placed in a fixture beneath the horn so that the runner is aligned with the long side of the horn (for horns with rectangular faces). The weld cycle is actuated in the normal manner, causing the horn to descend automatically and rest on the runner for a pre-set interval. Cycle times are generally less than 1 sec., including horn travel time (down and up). A 10-in.-long runner has been degated in a single energizing cycle at an average cycle time of 0.3 sec.

Horn pressure is kept light so as to impart maximum energy to the runner assembly. Normal pressure is 20 psig, but some variation up to a maximum of 40 psig may be necessary.

A sheet of resilient material, such as cork or urethane, is placed beneath the runner system to insure an out-of-phase relationship between part and horn. If this sheet is not used, the part vibrates in phase with the horn, and very little heat is generated. Urethane is preferred to cork, because cork eventually becomes flattened and loses its resilience.

Horn Design and Placement

Standard horns can be used for some degating jobs, but best results usually are obtained with a horn that is custom-made for use with a particular runner system. Contact between runner and horn should be maximized, so that energy is transferred efficiently.

Figure 11.1. Typical horn placement for balanced runner systems of same proportions but different over-all dimensions. (A) Parts can be removed from small-runner system with a single horn placement at main runner; (B) large-runner system requires two horn placements using a 10-in. horn; (C) four horn placements with a horn at least 8-in. long are needed with largest runner system.

Horn width should at least equal runner width. Flat-faced horns, which are normally carried in stock, are recommended for flat-topped runner systems such as those with rectangular or trapezoidal cross-sections. Round or oval runners require horns that are contoured to that shape. Some horns also have to be relieved in spots to accommodate bosses or other raised areas on the runners.

Horn length should ideally equal the length of the runner leg being degated, because plastics do not transmit ultrasonic energy well enough to permit degating at points over 3 to 4 in. from the horn extremities. Horns up to 10 in. long have been made, so many runner systems are small enough to be handled with a single placement of the horn. For large runner systems, or those with several branches (balanced runner systems), more than one horn placement may be necessary. This could be done automatically by using a sliding fixture that moves the runner relative to the horn. Alternatively, more than one horn can be used simultaneously, or a non-rectangular horn face can be used. For example, a circular horn has been used to remove parts from a radial runner system.

Either aluminum or titanium horns can be used. Aluminum costs half as much as titanium, but titanium horns are normally preferred, because they are three times stronger and will not fracture as readily when in use.

Small Gates Separate More Easily

In general, any gate cross-section that can fit into a 0.040-in.-diameter circle can be ultrasonically severed. Circular gates give the best results, because they melt uniformly. Some other geometries, consistent with the dimensional limitation, can also be used. Gates having a rectangular cross-section should not exceed 0.010 to 0.015 in. thickness in 0.040 in. widths. Heavier gates have been run successfully, but results were not reproducible enough for a production operation. Exceptions to the dimensional limitation are flash gates which work because they are so thin. A 1-in.-diameter radial flash gate has been successfully processed.

The gate size limitation places a limitation on part size. Larger parts in general require heavier gate sections to accommodate higher material flow rates. The largest part to be adequately degated thus far weighs 2 oz.; this part has a rectangular gate, ¼ in. by 0.030 in.

What It Costs

Both initial cost and operating costs depend heavily on the degree to which the process is automated. Automation requires a substantial investment in jigs, fixtures, and indexing equipment, but it can pay off in low operating costs for long runs. Automatic degating of non-identical parts from a common runner system requires some means of keeping the parts separate. The benefits of automation will otherwise be offset by sorting costs.

A basic ultrasonic welder (model 227 stand with J-17A power supply) costs about $2430. Horn cost depends on size, and ranges from $160 for a 2 in. by ¼ in. horn to $500 for a 10 in. by ¼ in. model. . . .

Chapter 12
Ultrasonic Soldering

Ultrasonics allows fluxless soldering and tinning of nonferrous materials with no chance of residual fluxes that could cause future corrosion and failure of the part. Soldering without flux of any kind means cost savings, since time-consuming fluxing and consequent post-cleaning are eliminated.

Ultrasonics enables the soldering of metals considered very difficult to solder properly, such as Kovar or nickel alloys used in the electronics and electrical industries. The soldering of materials considered by many as unsolderable in a production situation, such as aluminum wire used on transformers and motors, is easily accomplished with the addition of ultrasonics. Replacing copper wire with aluminum can offer great savings.

Gold embrittlement is virtually eliminated when ultrasonically soldering gold-plated materials. The ultrasonic energy disperses the gold plate within the molten solder and the brittle gold-lead-tin alloy layer is not produced.

The speed of the total soldering operation is greater because pre-cleaning and post-cleaning operations are usually eliminated. Also, actual soldering time is less with ultrasonics. Controlled thickness of solder coating can be produced because of uniform wetting of the surface.

How Ultrasonic Soldering Works

Standard 60-Hz electrical line power is converted to 20-kHz electrical power by a solid-state power supply. The power supply then delivers the 20-kHz electrical energy to a lead zirconate titanate transducer, which very efficiently converts the electrical energy into 20-kHz mechanical vibrations. The vibrations are transmitted through a specially designed horn. When the vibrating horn is immersed in a molten solder bath, the solder surrounding the tip cavitates, that is, bubbles form and collapse around it at 20,000 Hz. A part to be tinned is

Style	Description (Length x Dia. x Tip)
 Spade	.620 x .130 x .090
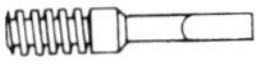 Screwdriver	.620 x .130 x .060
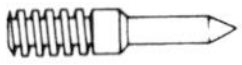 Pencil	.620 x .130 x .015
 Precision	.610 x .125 x .010

Plain Copper Tip

Style	Description (Length x Dia. x Tip)
Spade	.610 x .125 x .080
Screwdriver	.600 x .125 x .005
Pencil	.610 x .125 x .015
Precision	.610 x .125 x .005

Figure 12.1. Soldering tips (iron-clad, gold-plated copper tips).

then immersed in the area of cavitation around the horn tip, called the "active area." The cavitation of solder provides a scrubbing action at the interface of part and solder. Soils and oxides which prevent tinning are removed, leaving the part both mechanically and chemically clean. No new oxides can form on the part since no oxygen is present in the molten solder bath. This results in a lead which instantly accepts solder.

Ultrasonic Soldering Iron Equipment

Fluxless soldering is now possible through the application of ultrasonic vibrations to the tip of a soldering iron. The ultrasonics cavitates the molten

Figure 12.2. Typical ultrasonic soldering tool.

solder, which scrubs away corrosive oxides and soils. The surface of the workpiece is cleaned and receives a uniform solder coating without the use of flux.

The ultrasonic soldering iron solders aluminum, nickel and various other substances, as well as the more readily solderable metals. Good cohesive bonds can be made to ceramics, glass, plastic, fiber glass, quartz, and ferrites.

A typical ultrasonic soldering gun includes power supply, footswitch, and soldering iron. The power supply delivers 12 watts of ultrasonic energy at 20 kHz to the plug-in soldering iron when the footswitch is pressed, and also supplies electrical power for the 25-watt heating element in the tip of the iron.

A variety of tip inserts are available to expand the use of the ultrasonic soldering iron. An unheated tip can be used for larger parts, which are heated by an external source.

For micro applications, the heated soldering tips can be furnished in points of various lengths and diameters, down to needlepoint. The size of the solder ball that can be handled and deposited depends upon the techniques employed. Solder balls and solder microdots less than 0.010 in. can be deposited.

Specifications

Power Supply Dimensions:	Width:	8 in.
	Depth:	10 in.
	Height:	3¾ in.

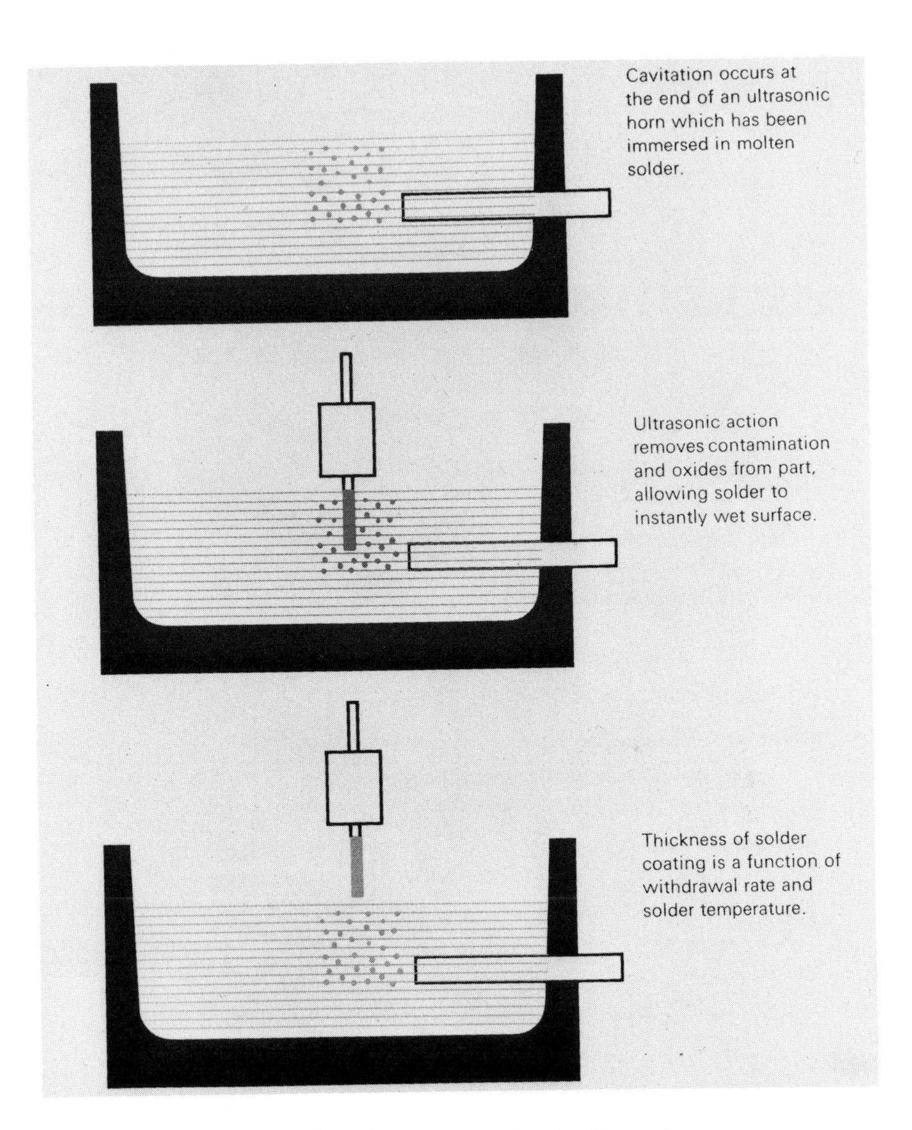

Figure 12.3. Tinning a part with ultrasonics.

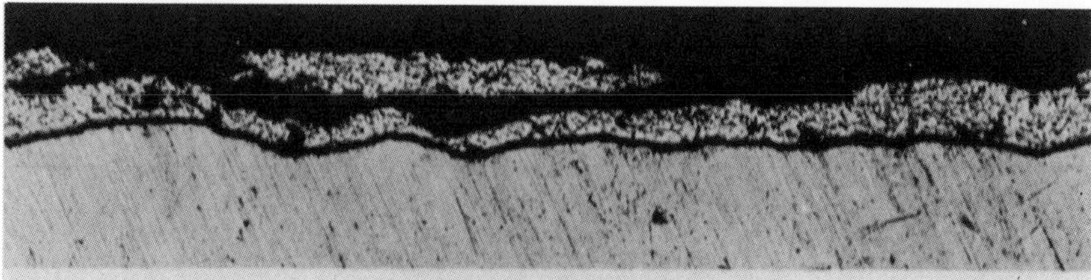

Without ultrasonics, considerable oxides are present at interface, resulting in poor bond. (Immersion @ 500° F for 1.5 seconds.)

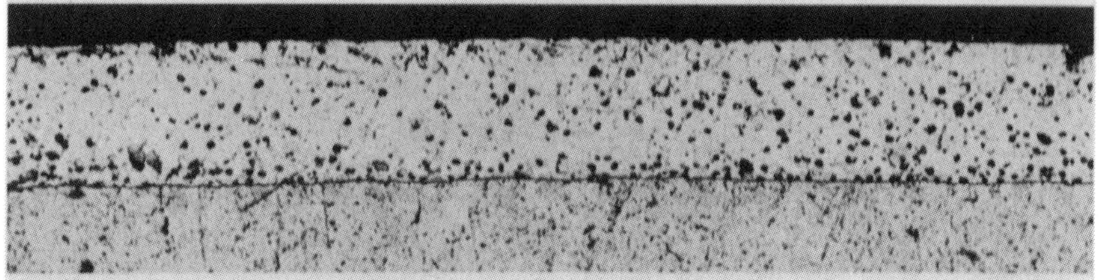

With ultrasonics, excellent bonding occurs since oxides have been scrubbed away as the solder wets the surface. (Immersion @ 500° F for 1.5 seconds.)

Figure 12.4. Bonding with and without ultrasonics

Line Power Requirements:	115 VAC 60 Hz, 1 amp.
Output Power – Ultrasonics	12 watts
– Heating element	25 watts
Controls:	Front panel heater control; tuning control; rear panel ultrasonic power control
Indicators:	Relative output power indicator; pilot light (incorporated in meter)

Application A

Objective:	To solder two aluminum tubes together at the flared joint area and create a mechanical coupling and hermetic seal (see Figure 12.6).
Base Material:	Aluminum tubing
Joining Alloy:	Zinc base solder

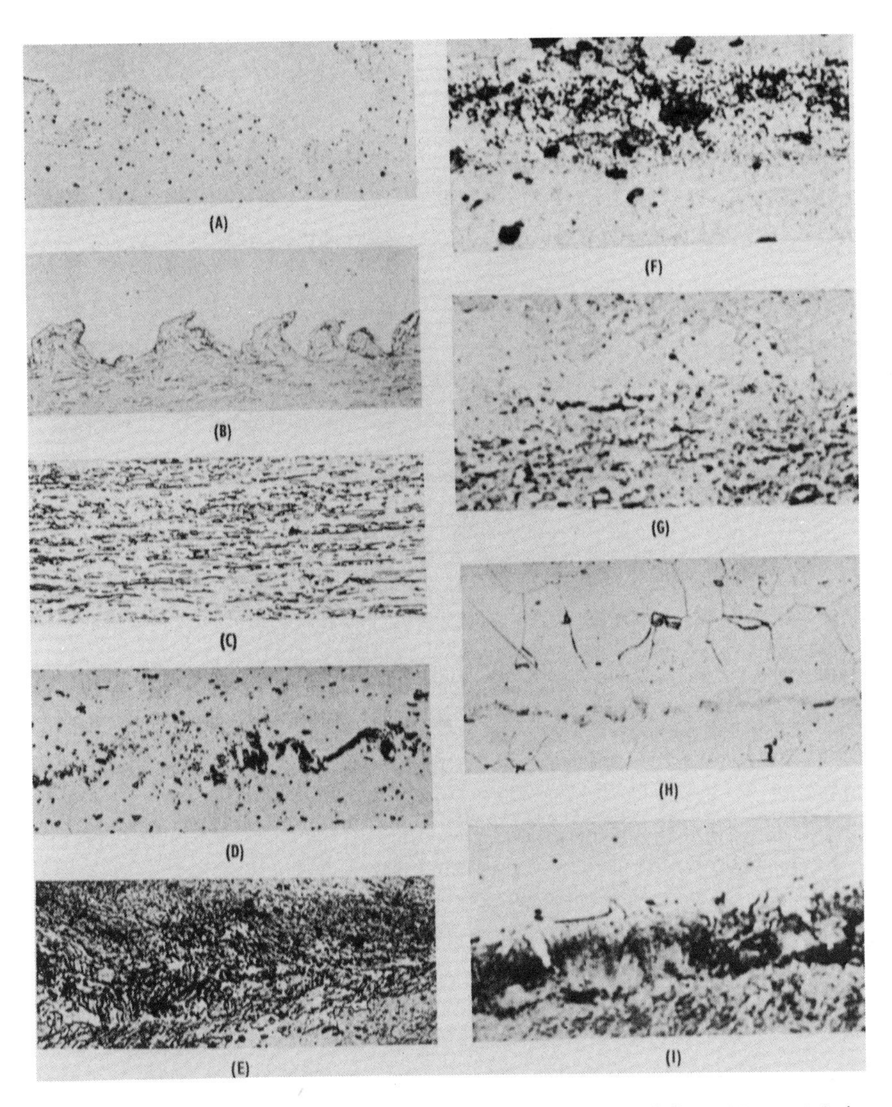

Figure 12.5. Photomicrographs of typical ultrasonic welds. (A) 0.003-in. nickel (top) to 0.003-in. gold-plated Kovar; (B) 0.005-in. nickel (top) to 0.020-in. molybdenum (X 200); (C) 0.008-in. arc-cast molybdenum to itself (X 140); (D) 0.012-in. 1100-H14 aluminum to itself (X 500); (E) 0.032-in. 2024-T3 bare aluminum alloy to itself (X 150); (F) 0.040-in. 2020 aluminum to itself (X 750); (G) 0.014-in. half-hard AISI 301 stainless steel to itself (X 500); (H) 0.012-in. solution heat-treated and aged inconel-X to itself (X 150); (I) 0.032-in. die steel (0.9%C) (top) to 0.032-in. ingot iron (X 1000).

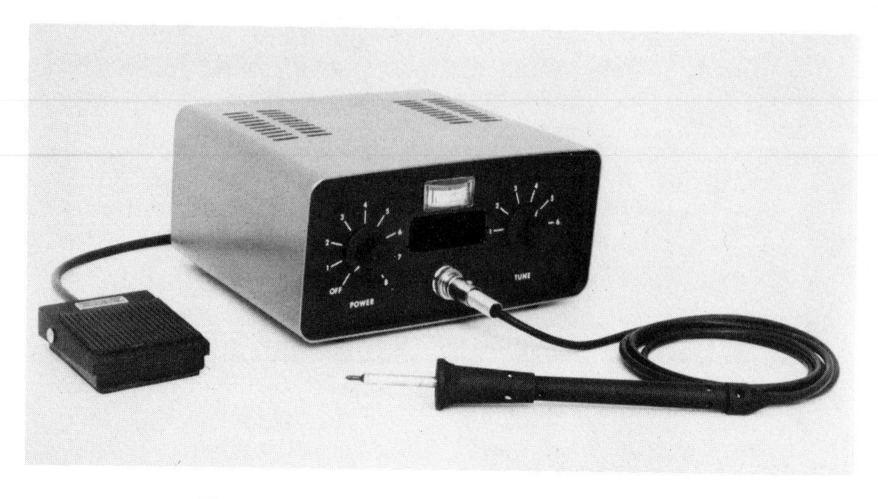

Figure 12.6. Portable ultrasonic soldering unit.

Equipment: Special ultrasonic soldering system

Solder Temperature: 700–800 degrees F

Soldering Process: The straight tube is inserted into the flared portion of the mating tube creating a partial seal. This seal is caused by an annular indentation in the flared section. A solder form is placed at the tip of the flared section. The assembly is heated in the joint area to liquify the solder. Once liquified, ultrasonic action is applied to the flared area to cavitate the liquid solder and allow it to flow down into the joint area. The solder remains in the joint area because of the annular ring. The cavitation action removes the oxides from the outside diameter of the straight tube and the inside diameter of the flared tube and creates a metallurgical bond, sealing the tubes together upon solidification.

Results: The use of ultrasonic soldering creates a mechanical joint and a hermetic seal at the flared section without the use of corrosive fluxes. This eliminates the post-cleaning required when flux is used.

Advantages: Ultrasonic soldering eliminates difficult cleaning problems due to flux use, and offers a more reliable product. Toxic fumes associated with flux soldering are eliminated.

Application B

Objective: Determine the feasibility of employing ultrasonic soldering processes to eliminate bridging effects and the formation of structurally brittle intermetallic compounds on gold-plated Kovar leads used in integrated circuits.

Base Material: Gold plated Ni/Fe

Joining Alloy: Tin/lead 60/40

Equipment: Special ultrasonic soldering system

Solder Temperature: 500 degrees F (or to customer specification)

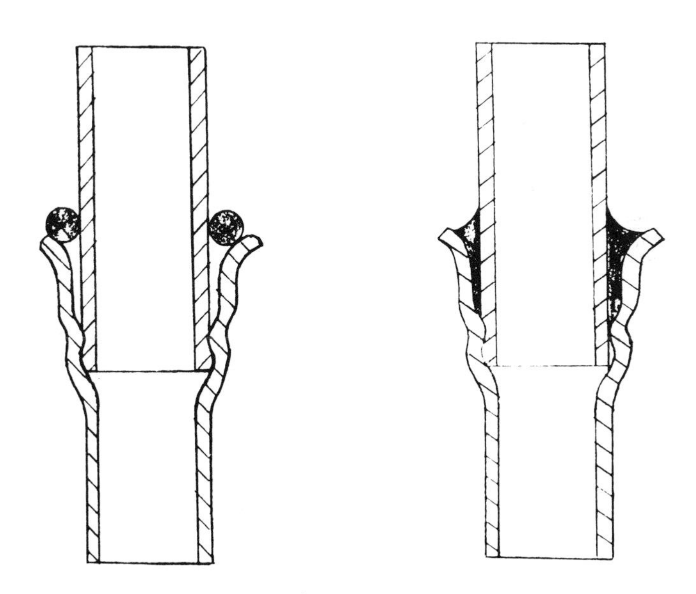

BEFORE ULTRASONICS AFTER ULTRASONICS

Figure 12.7. Illustration before and after ultrasonics.

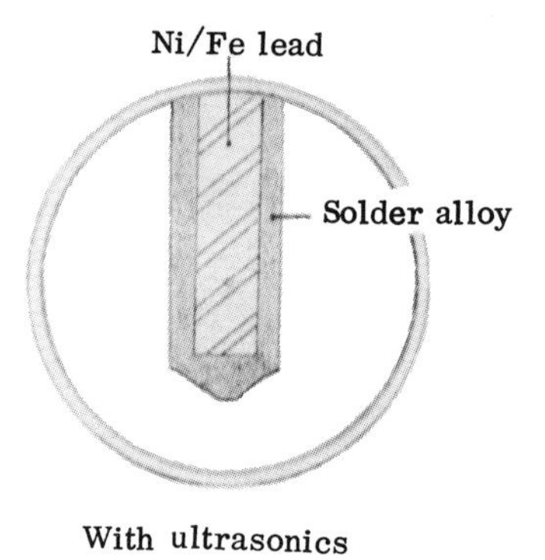

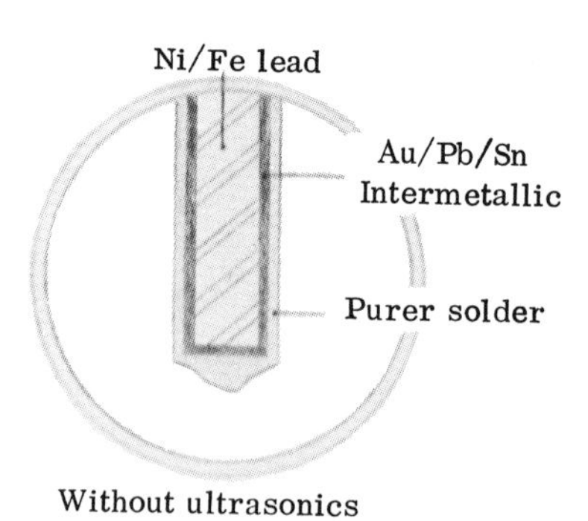

Figure 12.8. Welding structurally brittle intermetallic compounds.

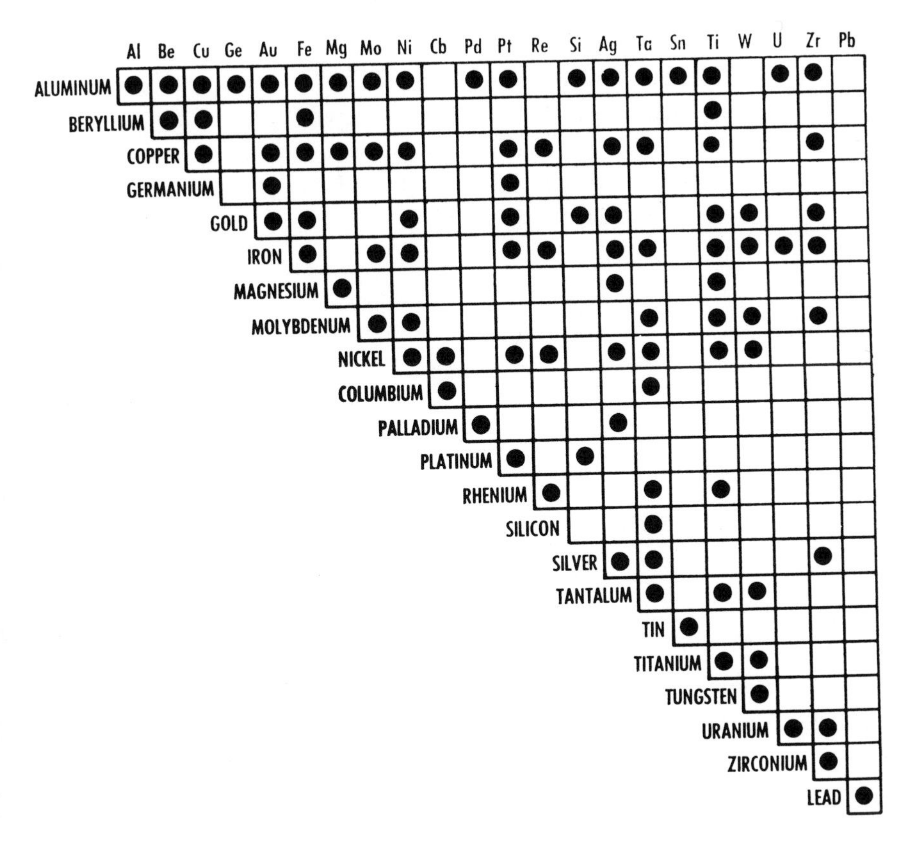

Figure 12.9. Metals and alloys that have been successfully joined by ultrasonic welding, or in which welding feasibility has been demonstrated.

Soldering Process: The sample was immersed centrally between the opposing ultrasonic horns by a vertical entry dripping operation through a dross-free surface for 1.0 second.

Results: The ultrasonically applied solder was found to be relatively free of deposits and smooth in appearance. There was no indication of oxide coating. This exposure assures oxide removal and uniform metallurgical bond integrity with complete gold-plating removal.

Advantages: Ultrasonic soldering eliminates the formation of the structurally brittle Au/Pb/Sn intermetallic compound layer. This facilitates better reflow characteristics of the leads in subsequent processes. Ultrasonic soldering action also lowers the surface tension and retards the bridging effects inherent in the dipping process.

Chapter 13

Ultrasonic Aerosolization

With 85 watt and 800,000 Hz "punch," by ultrasonics it is possible literally to break molecular bonds, rupture cells, and explode liquids into submicronic droplets.

It brings new dimensions to the arts of emulsification, particle dispersion and coagulation, aerosolization, turbulence creation, and the many chemical and physical processes based on rapidity and intimacy of particle mix.

The generator in the ultrasonic tool device utilizes a light-weight power unit that converts line voltage alternating current to a higher frequency of approximately 800,000 Hz. This high-frequency energy is conducted through an insulated flexible cable to a cylindrical piezoelectric transducer (1½ in. in diameter) which transforms the high-frequency electrical oscillations into mechanical vibrations (ultrasonic waves). When submerged in a liquid, the activated transducer produces a highly directional and powerful acoustic field, the total output of which can reach up to 85 watts.

The introduction of this sonic energy into a liquid system creates a veritable molecular storm. The energy is carried through the liquid by the oscillating motion of the molecules along the direction of propagation. This produces alternate adiabatic compressions and rarefactions, with resultant changes in density and temperature. In the typical case of an 800 kHz acoustic wave of high intensity traveling through water, the amplitude of motion of the molecules is small—on the order of 0.17 micron. But their acceleration attains values about 490,000 times greater than the acceleration of gravity.

Under such acceleration, the maximum instantaneous velocity reaches 86 centimeters per second, and the pressure amplitude at a given point in the water varies ($\Delta P = 12.75$ atmospheres) 800,000 times each second. This molecular turbulence results in a strong "radiation pressure" along the transducer axis, which creates a water geyser 7½ centimeters high. The internal turbulence within this geyser periodically disintegrates the water into tiny uniform droplets, with average diameters between 0.4 and 1.5 microns.

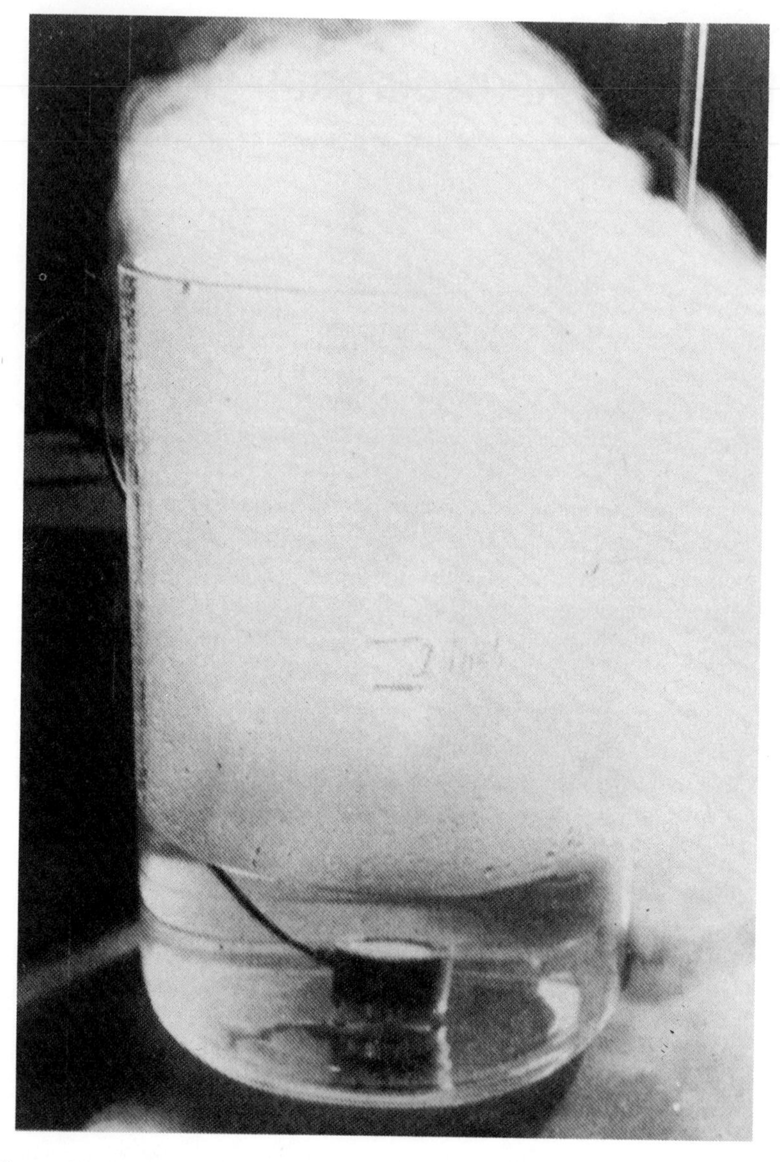

Figure 13.1. Submersible transducer during water aerosolization at a rate of 750 cc/hr. Average droplet size comprised between 0.4 and 1.5 microns.

It is easy to understand how such turbulence applied to liquid-liquid, liquid-solid, and liquid-gas mixtures can be made to accomplish useful functions. The ultrasonic "molecular scale storm" is great enough in many instances to sever molecular bonds, erode hard-to-dissolve particles, smash cells, and bring reactants together with energized impact.

Of tremendous importance is the fact that contrary to low-frequency irradiation (10–50 kHz), high-frequency insonotion does not release strong cavitation shockwaves throughout the liquid. It mainly acts at the molecular level by means of severe variations in acceleration. Since the acceleration amplitude is a square root function of the intensity, it means that continuous control of the magnitude of the phenomena observed at 800 kHz is possible by merely turning the output control knob of the generator.

Biochemists and microbiologists will greatly benefit from this continuous control feature which will enable them to gradually increase power while irradiating long chain molecules or cells. By doing so they will slowly affect molecule architecture or progressively extract cell components. This indeed is impossible at low frequency where brute cavitational forces are at once released above a certain threshold of energy and proceed in an uncontrolled manner.

In the past, high-frequency irradiation was limited by the use of expensive and fragile quartz-type transducers. The main difficulty with quartz lies in its high impedance, necessitating driving voltages up to 60 KV per centimeter of thickness. For instance, with an 800 kHz quartz of thickness $\lambda/2 \cong 0.35$ cm, a voltage of (60) (0.35) = 21 KV is required. This requires a complicated crystal mounting and insulation to carefully avoid flashover from front surface to back surface around the crystal edge. The new high-power piezoceramics used do not present such inconvenience since they need approximately 100 volts RMS to be driven at peak output.

Until recently, piezoceramic elements could not be driven at maximum power over extended periods of time in most chemical solutions. This is understandable because the thin conducting silver layer at the transducer upper face was not able to resist the huge accelerations produced in the high-frequency region. Even gold or platinum (0.25 in. thick) nonporous metal films would experience severe erosion after a few hours at peak excitation. Indeed, as soon as minute openings are created in the upper face protective layer, chemicals can penetrate inside the ceramic and provoke its partial depolarization, which in turn will initiate heat and dielectric losses.

In order to cope with this problem, ultrasonic equipment manufacturers found, after exhaustive research, a type of glass-like material that can be sprayed over the vibrating face of the barium titanate and protect it permanently against mechanical or chemical corrosion. It enables long-lasting, continuous, high-intensity irradiation (more than five thousand hours) at liquid temperature in the 145 to 212 degrees F range (63 to 100 degrees C).

The generators are so powerful that they enable irradiation of test samples through thin glass, mylar membranes, or thin metal containers. In this case water can be used both as a coupling medium and efficient heat sink. A slight loss of energy will be sustained during passage of the acoustic energy through the container walls.

Applications

- Emulsification of multiphase liquid systems
- Particle dispersion in fluids
- Particle coagulation
- Atomization in flame spectrophotometry
- Acceleration of chemical reactions
- Wind tunnel visualization
- Accelerated cleaning

Colloidal dispersions of rocket propellants

- Dispersion of catalysts or reagents
- Radioactive fog tracing
- Aging alcohols, wines and spirits
- Catalyst activation or crystallization
- Filtration acceleration
- Biological and medical extractions
- Cell destruction—sterilization
- Dispersion of anesthetics, perfumes

- Aerosol therapy—drug inhalation
- Humidification in homes, offices, hospitals
- Control of static charges in surgery, testing and clean rooms
- Impregnation, coating or dyeing

Chapter 14
Ultrasonics in Plastics Fabrication

The best understanding of the broad utility of ultrasonics in plastics assembly can be most quickly developed by citing specific applications where ultrasonics have been used to solve particular assembly problems.

Using an Ultrasonic Sealer for Hermetic Sealing of Vials of Caustic Solution

Problem: A battery company wanted to offer its customers a long battery shelf-life. It couldn't offer the 10-year or over shelf-life because it could never be sure how long the seal of the vial which contained the battery's caustic solution would last. This was, traditionally, the weakest part of the battery.

The caustic will leak through even the tiniest fissure in a seal. The vials had always been sealed with solvent. But solvent sealing left too many air bubbles, and the caustic forged a leakage path through them.

Solution: Ultrasonic sealing was a more desirable technique, but an impressive array of technical problems immediately presented themselves (Figure 14.1).

First, the vials had to be sealed with the caustic in them. Any energy applied to an acrylic vial of this type was bound to generate cavitation—or the violent agitation of the fluid in the vial. Since agitating caustic was inherently dangerous, a safe method had to be discovered to dampen the cavitation.

Second, the cap of the vial had to be hermetically sealed all around. To do this in the conventional way required precise positioning of the cap in the vial when the energy was to be applied. If the cap fit too loosely or too deeply in the vial, a hermetic seal could not be assured.

The solution was to design a 0.002-in. radius bead around the inside of

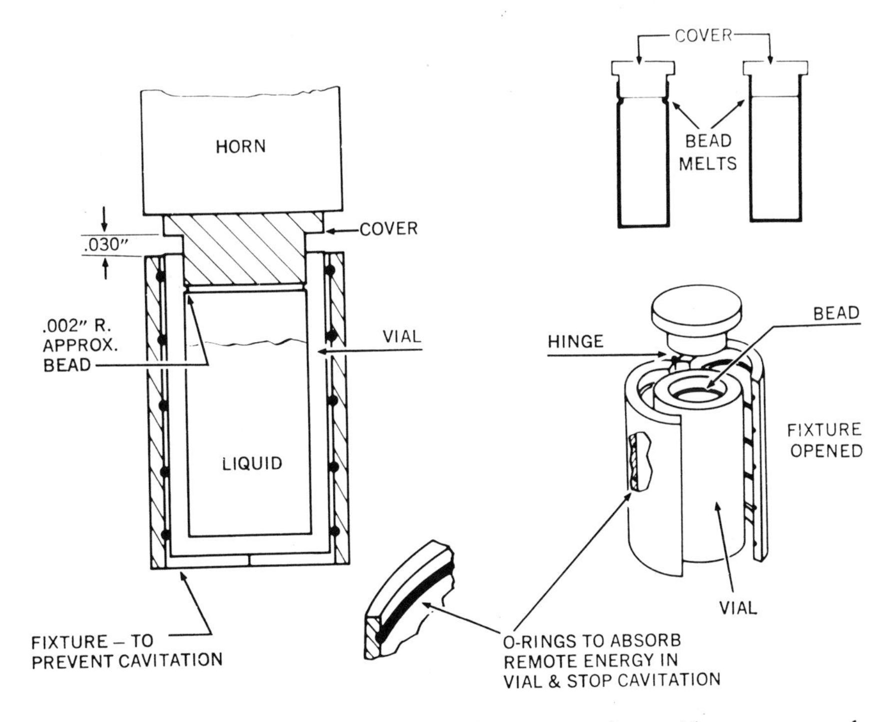

Figure 14.1. Ultrasonically sealed vials for battery caustic provides a secure package until ready for use.

the vial. The bead was made at exactly the point where it was most desirable for the cap to sit in the vial before sealing. The operator thus had a foolproof method for positioning the cap.

When the ultrasonic horn was applied to the cap, energy flowed straight down, so the bead actually melted and bonded itself to the cap as it was forced down. The result was a perfect hermetic seal each time.

The engineers involved designed a unique fixture to prevent cavitation. The same circular shape as the vial, the fixture clamped around the vial to hold it in place under the horn of the sealer. A number of rubber O-rings were built into the metal wall of the fixture. These rubber rings severely dampened the energy flow so the caustic is never agitated enough to interfere with the sealing process.

Using an Ultrasonic Seal for No-Reject Polystyrene Spools

Problem: A typewriter spool is an example of a plastic part, minor to the overall scheme of the machines of which it is a part, but which could be an annoying cause of trouble when it breaks in use. A typewriter manufacturer has been sealing the two halves of the spools with solvent, and the seal in a number of them had cracked when the typewriter was in use.

The solvent used also created unpleasant fumes and was a chronic source of complaints from the production staff.

Solution: A group of engineers helped design the sealing operation shown in Figure 14.2. A machined fit produces a lock between the skirt and the hub

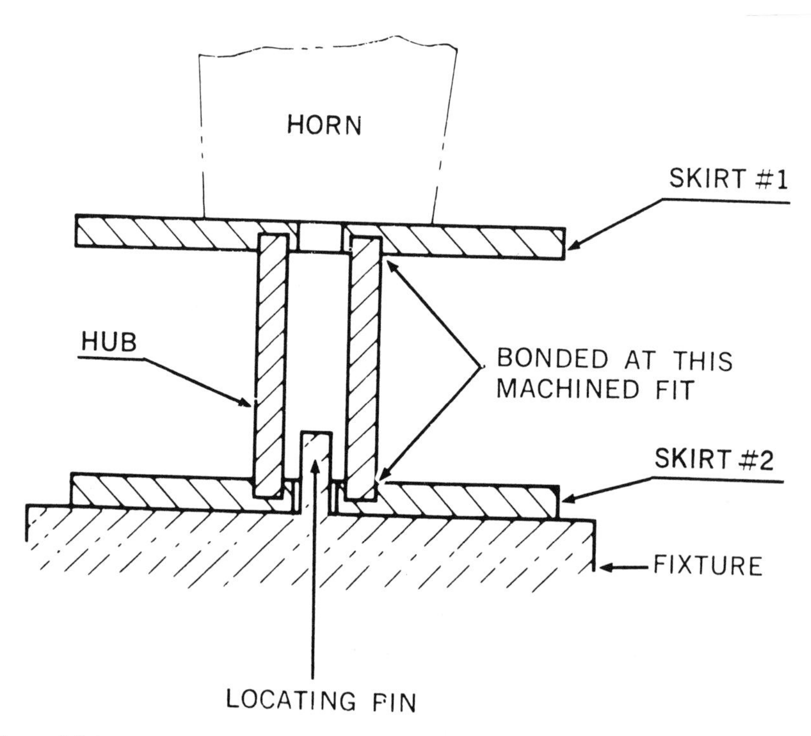

Figure 14.2. Typewriter spool assembled ultrasonically provides a stronger joint than solvent-cemented construction.

of the spool. This fit allows the bond to be effected at both skirts, automatically, with a single application of energy.

The mechanical fit allows the parts to be in perfect alignment for the strike each time. Otherwise, it would have been necessary to devise a fixture to hold the parts in position.

An ultrasonic seal is the most perfect bond known for this type of polystyrene. In this case, failures in the field stopped completely when the parts were sealed ultrasonically.

Ultrasonics Reaches Inaccessible Sealing Area

Problem: A wig-case manufacturer recently decided to ultrasonically seal the hinge joining the two parts of his case. Two factors affected his decision: (1) the economics of using this high-speed method, and (2) the fact that he needed to join the two parts without marring the appearance of the case (Figure 14.3).

Solution: To avoid marring the case, the ultrasonic sealing energy had to reach the hinge from the inside of the case so that there was no possibility of flash on the outside finish.

By ultrasonic sealing, the manufacturer was able to meet both the aesthetic and economical requirements by designing a special tool. The generator in an ultrasonic sealing machine produces electrical energy. The tool transmits this energy to the part to be sealed. In this particular application the tool had to go all the way through the large wig case to get to the inside hinge. The engineers designed a full wave-length 10-in. tool, instead of the conventional half wave-length tool, to transmit the energy to the sealing area. In addition to its length, the tool has five prongs, 1/16-in. diameter each, at its output end. These set the tool securely in the polypropylene, and enable the ultrasonic energy to penetrate through the heavy plastic to the hinge where the seal must be effected.

Using an Ultrasonic Sealer to Seal an Acrylic Reflector and Lens Assembly

Problem: The reflector inside the lens housing had to be fabricated in a tail-light assembly.

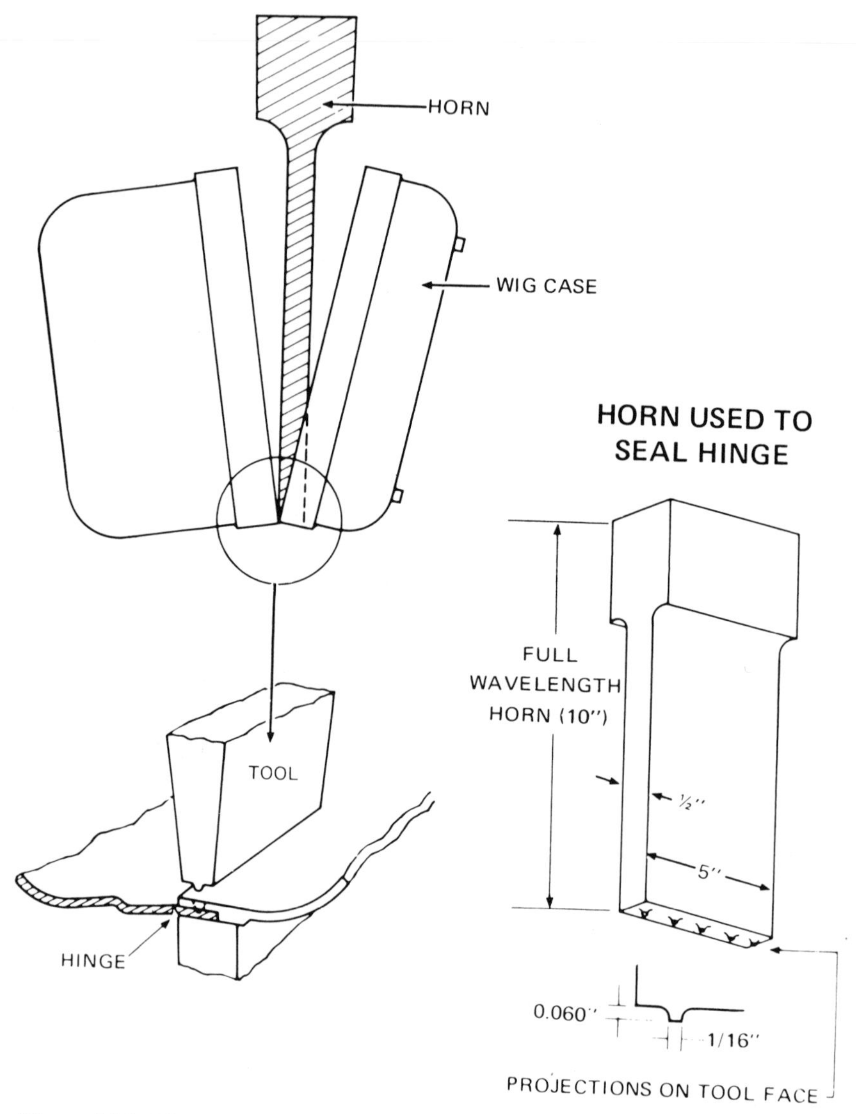

Figure 14.3. Ultrasonic sealing of wig-case halves provided improved economics and gave a marr-free joint.

Heat staking with projections was attempted, but this produced immediate difficulties. The scrap factor reached an unacceptable level of 10% and could not be reduced. The job required skilled labor and ideal conditions.

Even among those assemblies successfully sealed, the acrylic had a tendency to embrittle at the stake and rattle itself loose.

Solution: In order to seal the assembly ultrasonically, it was necessary to contact seal at two ends at the same time. A slotted tool was designed for this purpose (Figure 14.4).

The slotting in the 4-in.-wide tool concentrated energy equally at its two ends, and both sides were sealed simultaneously.

The scrap factor was slashed from 10% to virtually zero. The ultrasonic seal did not loosen, or embrittle the acrylic, so the assemblies stayed sealed.

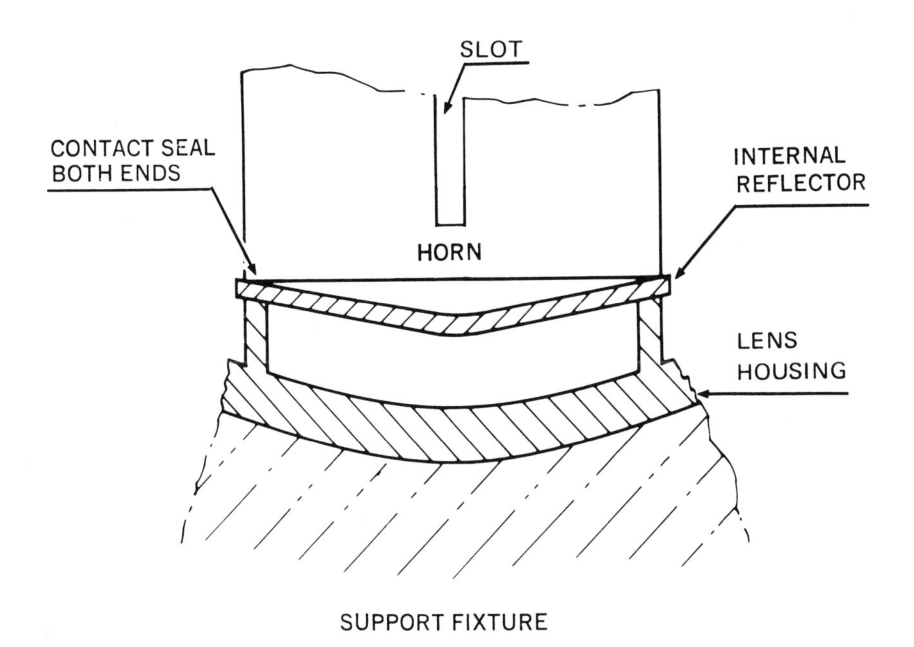

Figure 14.4. Slotted tool concentrates ultrasonic energy equally at its two ends to seal internal reflector to lens housing.

Sealing Polyurethane Foam around a Metal Part

Understanding the nature of various kinds of materials is as crucial to solving ultrasonic sealing problems as understanding the nature of ultrasonic sealing itself.

Problem: A good case in point was the problem faced recently by a manufacturer of a part used in automobiles for smog control. The part consists of a metal filter element which is surrounded by polyurethane foam. The foam had to be bonded so that fuel would not penetrate the seal, yet would allow the metal filter element to nestle tightly inside (Figure 14.5).

Solution: The manufacturer developed a special anvil on which the foam could be seated with the metal filter placed on top. However, when any kind of a die was brought down to the anvil, the metal of the anvil itself conducted the heat away and failed to effect a tight seal.

A Teflon ring was put into the metal anvil at the point where the foam was to be sealed. When the horn was brought down (see Figure 14.5) so that the two halves of the foam were touching each other and the ultrasonic energy flowed through them, the Teflon retained the heat energy. Teflon was selected because it is such a poor thermal conductor, and can withstand high temperatures without distorting.

Water channels were also built into the anvil to keep it cool enough for the operators to touch.

Ultrasonically Sealing Tape Cassettes

Problem: When a manufacturer of tape cassettes decided to use ultrasonic sealing, one of the most dramatic results was slicing the labor force for the welding operation right in half.

The cassettes had been sealed with five screws which were inserted by punch presses. But the punch-press screwing operation was full of problems. The screws sometimes went in askew, sometimes were too soft and frequently held up production. In fact, the two time-consuming operations involved in sealing the cassettes took nearly as long as assembling the internal parts.

Solution: An ultrasonic sealing system changed all that. The system

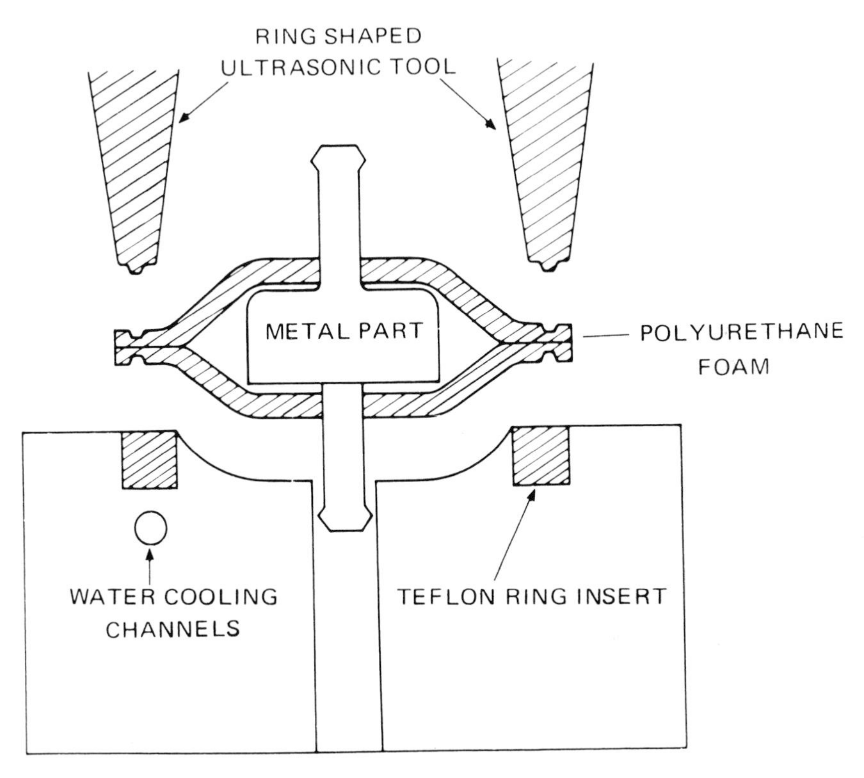

Figure 14.5. Teflon insert in anvil portion of assembly tool prevents heat from being conducted away from joint being sealed.

designed truly automated not just the welding step, but the delivery of the parts to the machine and their removal.

One of the advantages of ultrasonic sealing is that it is usually people-proof and goof-proof. The tape cassette manufacturer found, after twleve months of operation, that rejects from the sealing operation were 100% eliminated.

In terms of quality control, there just isn't anything, including the oldest, simplest methods known, that approaches the reliability of ultrasonic sealing.

Thermal Cup Welding

Problem: Thermal cup welding is generally considered an excellent application for ultrasonic welding of plastics. Joint design plays an important part in producing a weld that can withstand physical and thermal abuse such as experienced in a dishwasher. There must be an hermetic seal between inner and outer shells to make the thermal principle effective. If correct joint design is followed, hermetic seals are possible.

Solution: The insert in Figure 14.6 illustrates the location and approximate proportions of the suggested joint. Here, the energy director is placed on the inside step, to produce a weld without flashing to the outside. Because of the many different sizes and shapes of cups, each one has to be considered separately.

The ideal position for welding places the horn closest to the joint area; however, this may cause marking on the upper edge due to the small contact area. In such cases, depending upon the material, cups may be welded in an inverted position to provide better coupling and transferral of ultrasonic energy to the joint area.

Ultrasonic Tamper-proofing in Methadone Program

Problem: The New York State Narcotics Addiction Agency has approved the use of ultrasonic welding equipment for a concept developed by KLM Corporation of Stratford, Conn. The system is used to tamper-proof polystyrene containers used in a new methadone treatment program (Figure 14.7).

Solution: Three years ago, the KLM Corporation successfully developed and patented a high-speed, low-cost method of producing plastic caps to replace more expensive metal caps for glass and plastic containers. The recent swing to plastic bottles inspired the company to experiment with ways to prevent cap back-off, or loosening, a major problem in the closure industry. Back-off is caused by vibrations during shipment, or as a result of expansion and shrinkage because of temperature changes.

KLM had some knowledge of ultrasonics, and believed the process might be used to seal the plastic caps, eliminating back-off—and as a bonus, tamper-proof the container at the same time.

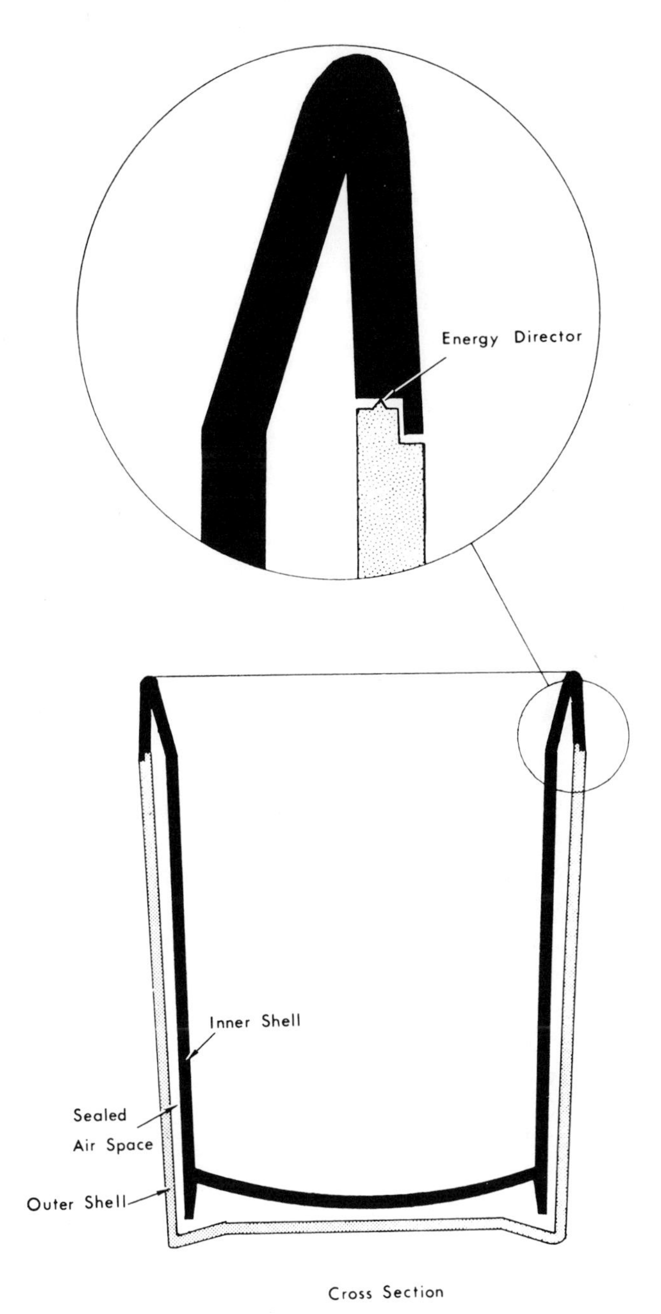

Figure 14.6. Energy director in thermal cup joint produces strong weld without flashing to the outside of the joint.

Figure 14.7. Tamper-proof seal prevents back-off as well as assuring that the contents remain fresh.

The outcome of the company's research is a seal produced in a fraction of a second. The metal at the skirt of the cap is deformed to tack around the transfer head on the neck of the bottle. The strength of the tack weld is sufficient to prevent back-off, but the cap can be removed as easily as conventional aluminum caps. The distinct pop or snap heard when the seal is broken is proof that the contents of a sealed container are still fresh.

Ultrasonically Reactivated Adhesives Bond Copper and Wood

Problem: Reactivation of adhesives is one of the less common uses for ultrasonic energy. Nevertheless, Wire Noveltys Manufacturing Company of Shelton, Conn., has reduced production costs at least 50% on high-quality carpenters' squares they are manufacturing for General Hardware Manufacturing Company and Sears (Figure 14.8).

Solution: The handle of the carpenters' square is resin-impregnated rosewood with copper trim and a steel scale. The wood handle is set into a metal

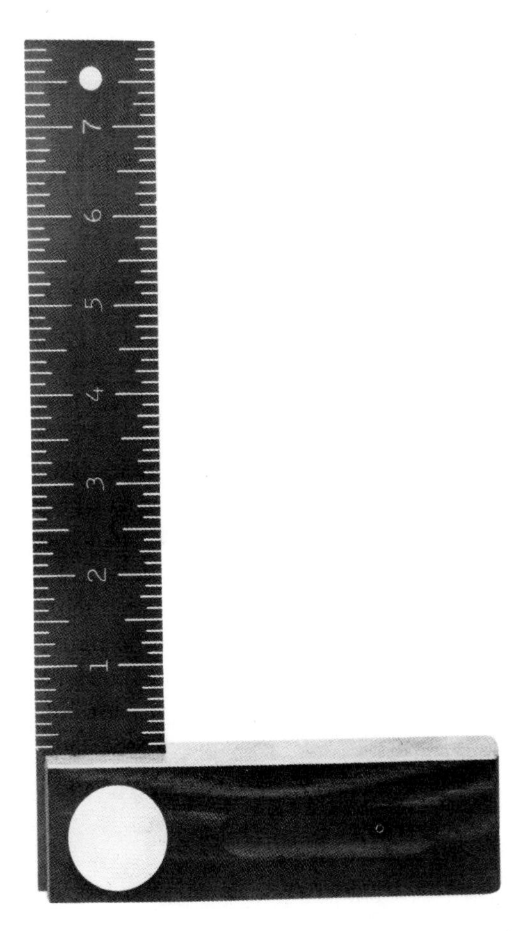

Figure 14.8. Adhesive, activated by ultrasonics, bonds copper trim to wooden handle of carpenter's square.

fixture. A strip of plastic film adhesive is positioned between the wood and a strip of decorative copper. A slotted bar horn contacts the copper for 1.5 seconds of ultrasonic exposure to activate the adhesive. A 3-second hold time allows the adhesive to set. A slot for the scale is cut into one end of the handle. A jig provides a perfect square condition while two pins are press fit into pre-drilled holes to lock the scale in place. Two circular pieces of copper are placed over the pins and held in place with adhesive, which is also ultrasonically reactivated by a 1-in. catenoidal horn. After sanding, the handle is given a celar lacquer finish.

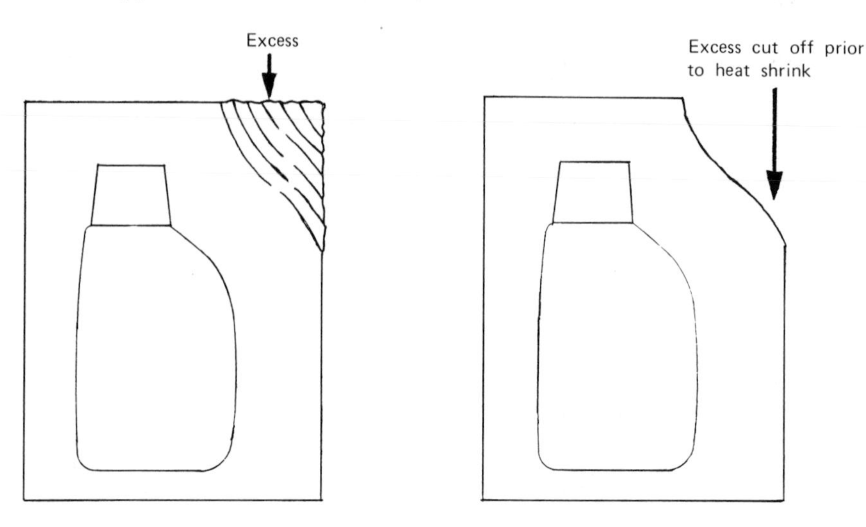

Figure 14.9. Ultrasonic cutting.

Ultrasonic Shrink Packaging

Sea & Ski conducted a market survey of three types of packaging: shadow box, carton, and unwrapped plastic bottles. The unwrapped form was most successful because the familiar slope-shoulder bottle was easily identified by the customer. Even so, the naked bottle had its drawbacks. Sea & Ski sun care products have a five-year shelf life, but only a ten-week buying season. By the time retailers returned unsold merchandise for salvage, bottles had been price-stamped with ink and gathered dust through static electricity.

Clear-film sealing was a logical answer. Pricing and dust could be removed simply by tearing off the film. In addition, shrink packaging cost only 0.003¢ per bottle, versus shadow boxes costing 0.0125¢, or cartons at 0.0225¢.

Slope-Shoulder Bottle. Bottles for sample film wrap were submitted to Oliver Machinery Company, one of the largest manufacturers of automatic packaging equipment. When sealed by shrinking in standard rectangular PVC bags (Figure 14.9), an unsightly blob of excess material collected in the shoulder area of the bottle. Solution—cut off excess material prior to heat shrinking.

Existing cutting and sealing equipment did not work. Hot wire or rods caused film misalignment because of excess drag. In addition, direct contact be-

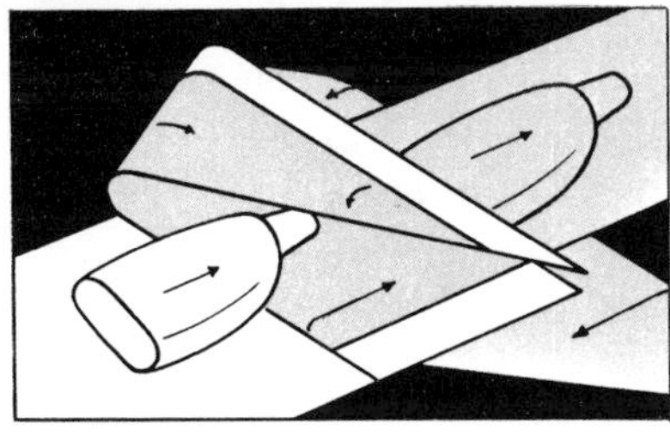

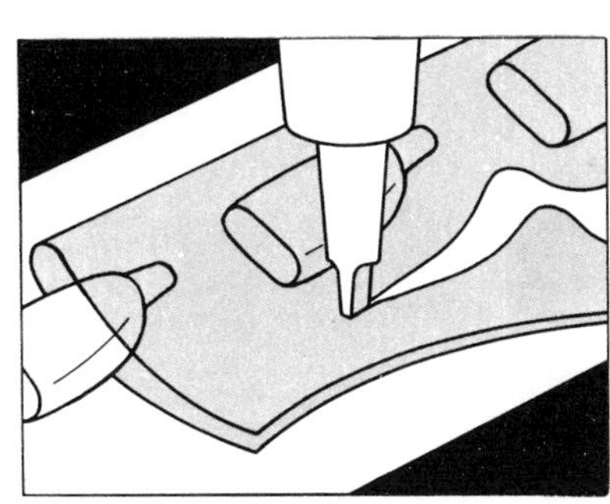

1. Bottles are manually loaded on evenly spaced pusher lugs in the infeed conveyor system.
2. Forming head takes flat 75-gauge PVC film and forms it into a horizontal tube. Bottle is inserted between the center folded film via the in-line delivery system.
3. Film is contoured to the shape of the bottle and simultaneously sealed and trimmed as it passes under the ultrasonic tool at rates of 120 bottles per minute.
4. Radiant impulse wires burn off and seal bottles in individual pouches. Contoured PVC bag as it appears prior to entering the shrink tunnel.

Figure 14.10. Ultrasonic sealing: plastic film to paper.

tween PVC and a constant heat source decomposed the material, releasing noxious chlorine gas and carbon residue. The gas caused a danger to health and safety, and rusting of machinery. Carbon particles collected on sealing elements and had to be scraped off. When hot wires were operated at reduced temperatures to minimize charring, film stuck to the elements and pulled away in fine strings called "angel hair," requiring hand trimming.

Ultrasonic Contouring. Oliver discovered that ultrasonic cutting and sealing eliminated PVC degradation and its by-products, carbon build-up and chlorine gas, because the ultrasonic tool remained cool. Its mechanical vibrations at 20,000 cycles per second are transmitted to the interface of the material where they melt, fuse and cut the film. Horn pressure, power and amplitude are precisely controlled so material reaches temperatures at the low end of PVC's melt spectrum. These temperatures are too low to produce material degradation, yet sufficient to prevent formation of angel hair.

To make the ultrasonic equipment mobile, the J-converter, catenoidal horn and cutting anvil were mounted on a cam-operated driveshaft. The shape of the cam was designed to move the ultrasonic tool around the slope shoulder of the bottle. PVC film was shaped, sealed and trimmed in a continuous operation as each bottle traveled to the shrink tunnel. The system is capable of handling 120 bottles per minute. Two-, four-, and eight-ounce bottles can be contoured by merely changing the cam mounting location. The system can also make continuous straight trim seals by removing the cam (see Figure 14.10).

Chapter 15

Ultrasonic Sewing

A new sonic sewing machine that integrates the sonic sewing head into a standard industrial sewing table has been developed. Materials with up to 35% natural fiber content are fed between the vibrating tip of the horn and a specially designed anvil, called a stitching wheel, which creates the stitch pattern. Feed rates are controlled by the operator and range from slow speed for detailed work

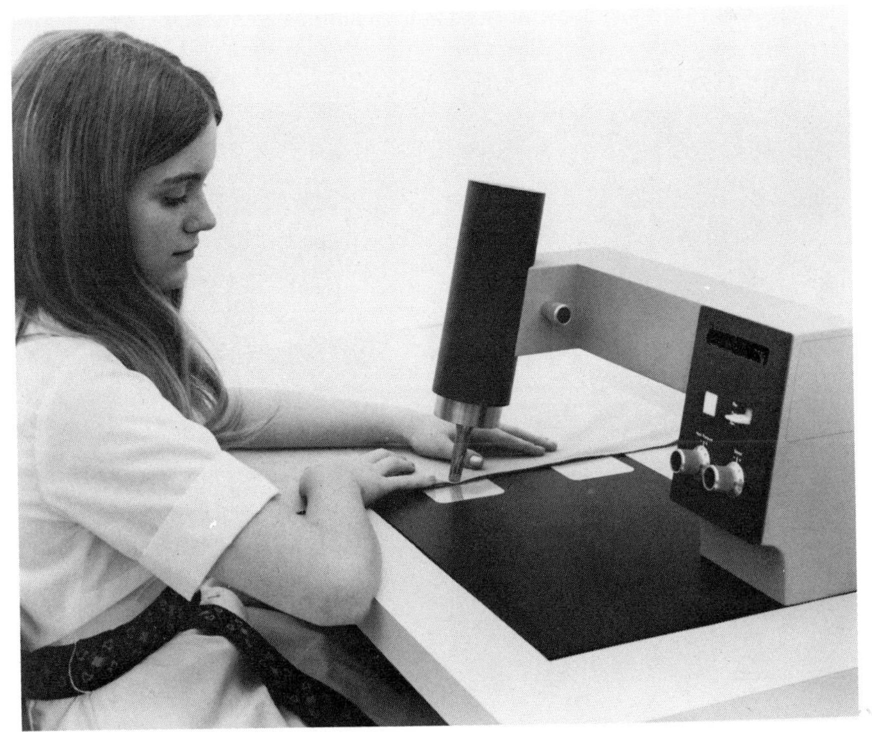

Figure 15.1. Ultrasonic sewing eliminates the problems associated with needles, thread, and bobbins of the conventional machine.

to 50 feet per minute. The sonic sewing machine can also trim and seal edges simultaneously.

Latest Equipment

Ultrasonic energy is generated by a solid-state power supply, converted to mechanical vibrations and channeled into the sewing tool or horn (Figure 15.1). When materials are fed between the vibrating horn and a specially designed stitching wheel, frictional heat created by the horn bonds the material in the pattern of the chosen stitching wheel. Wheels, available in a wide variety of patterns, can be changed in as little as 20 seconds (Figures 15.2–15.5).

Since ultrasonic sewing eliminates thread, needle and bobbins, there is no color matching, threading, downtime for breakage of either thread or needle, bobbin rewinding or inventory. Elaborate sensing devices used to detect broken thread, skipped stitches and empty bobbins in automated systems are eliminated.

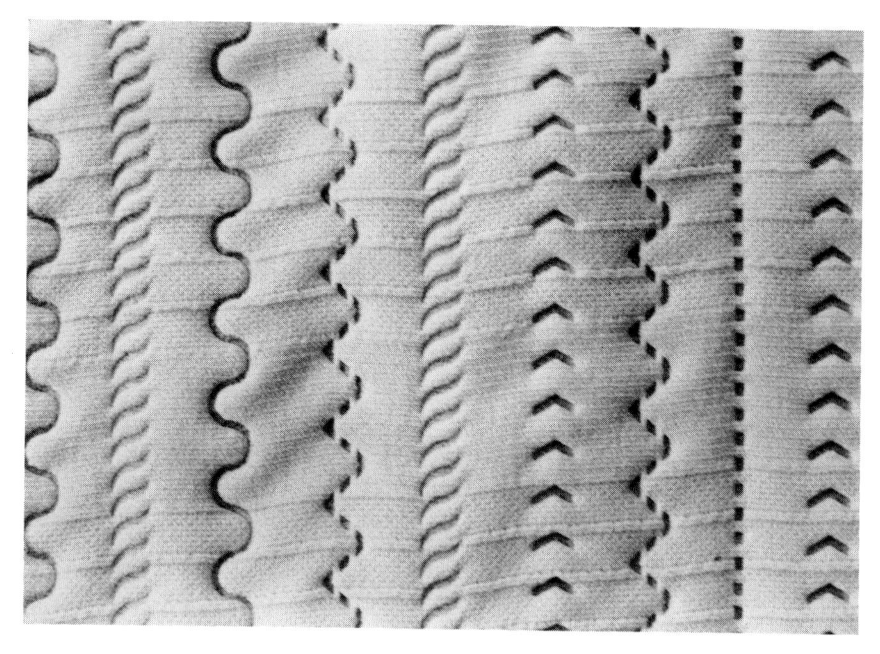

Figure 15.2. Ultrasonics fuses synthetic fabrics, containing up to 35% natural fibers, to a permanent bond.

Figure 15.3. A variety of patterns is available by using different stitching wheels that can be changed in 20 seconds.

A comfort-engineered treadle controls sewing speeds from extra slow for detailed work, to full speed (up to 50 feet per minute) for effortless maximum output. Operator training and error are minimized because of simple, easy-access operating controls mounted at eye level (Figure 15.6).

The offset sewing arm is positioned to allow maximum unobstructed work space for better set-up and maneuverability of material. The sewing area is plainly visible because the head is slanted 12 degrees away from the operator. Ample space is provided for mounting accessories. The newer models have pre-engineered power takeoff points for attaching devices such as puller and roller feeds.

There are no eccentric or reciprocating parts to oil, and little heat is generated because the machines operate at less than 100 rpm, even while sewing at 50 feet per minute.

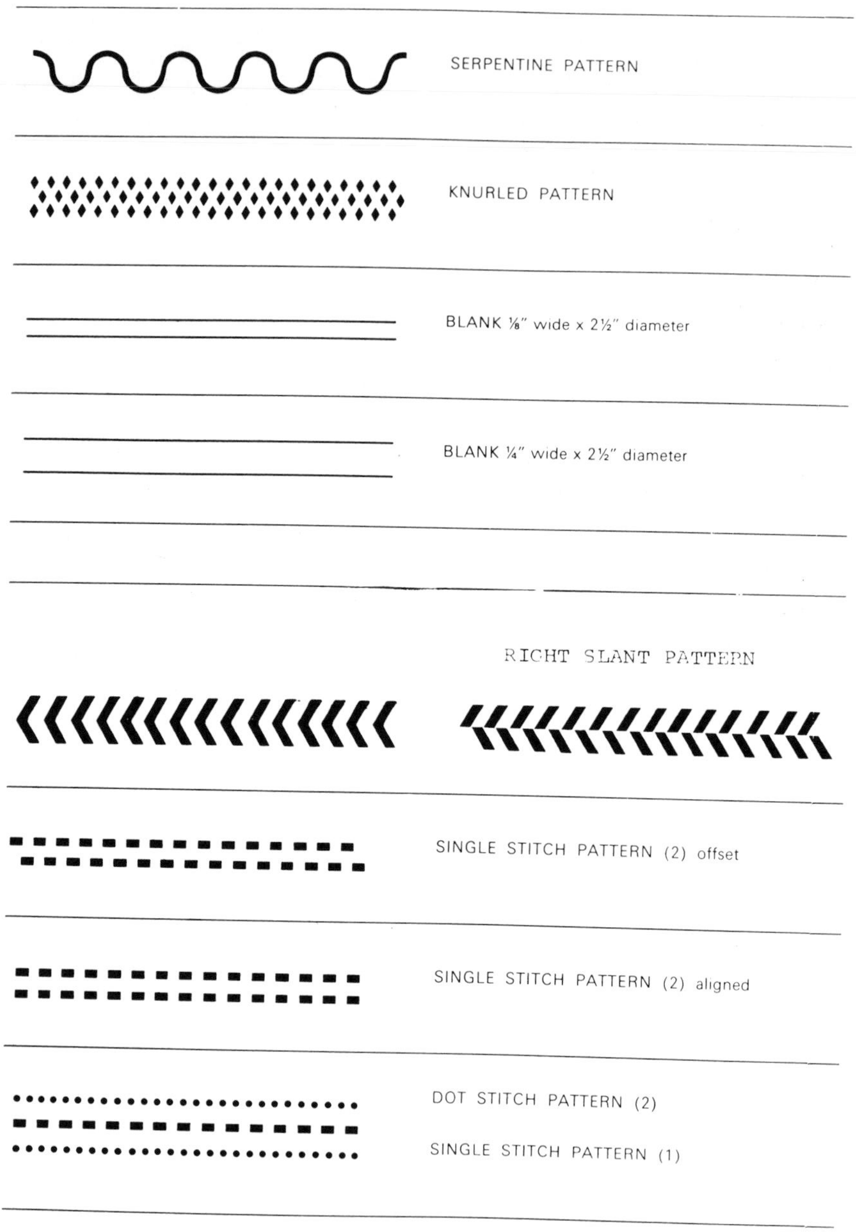

SERPENTINE PATTERN

KNURLED PATTERN

BLANK ⅛" wide x 2½" diameter

BLANK ¼" wide x 2½" diameter

RIGHT SLANT PATTERN

SINGLE STITCH PATTERN (2) offset

SINGLE STITCH PATTERN (2) aligned

DOT STITCH PATTERN (2)

SINGLE STITCH PATTERN (1)

Figure 15.4. Ultrasonic base stitching pattern.

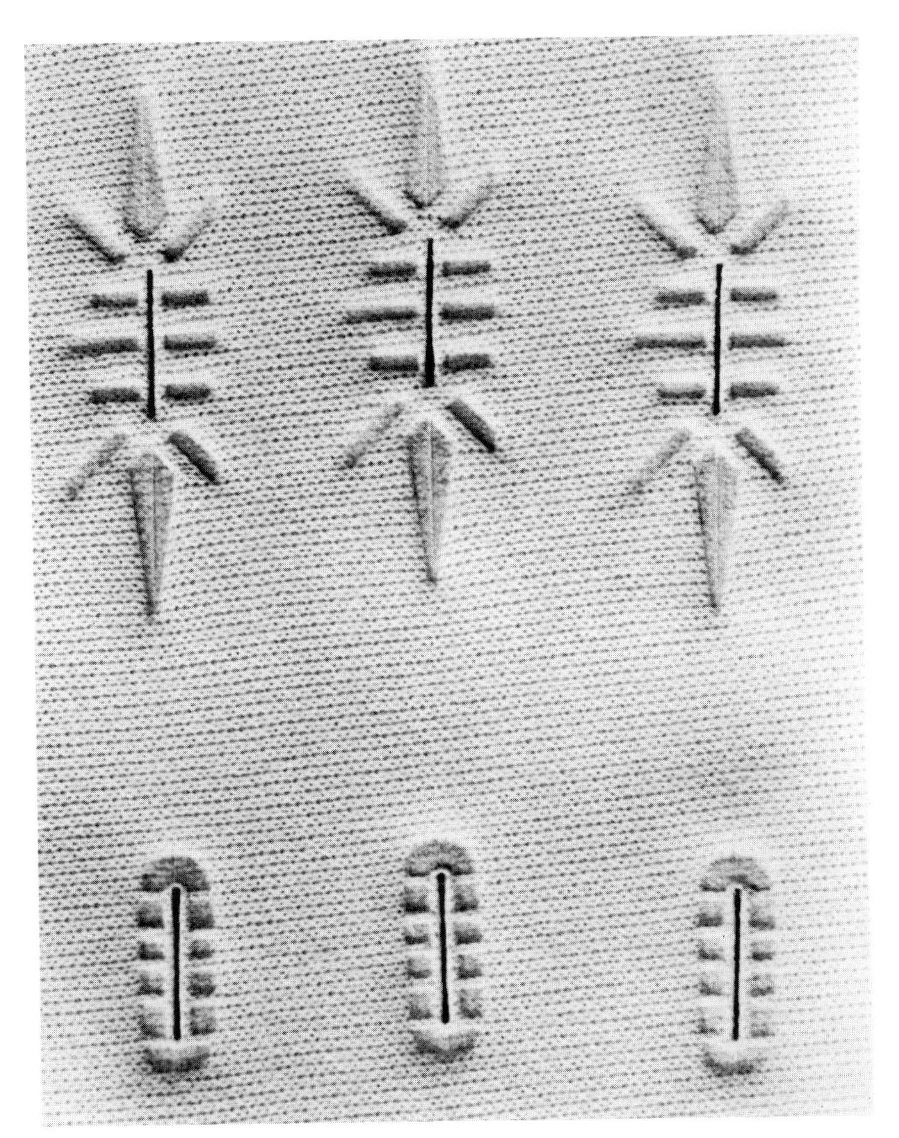

Figure 15.5. Embossed and slitted stitching.

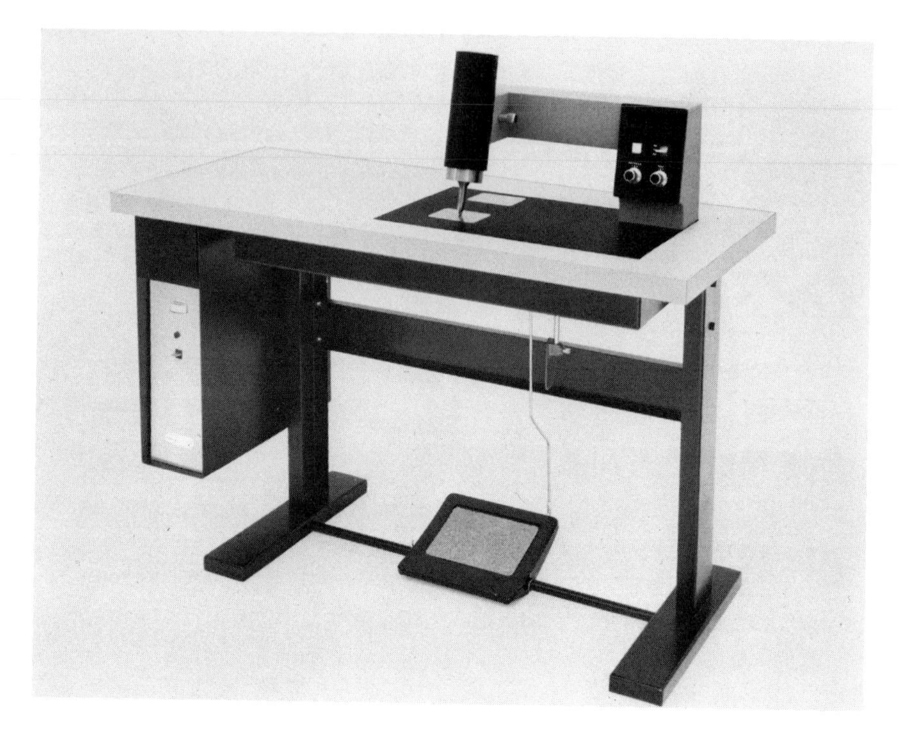

Figure 15.6. Typical ultrasonic stitching machine.

Accessories

A broad range of standard "stitch" patterns is available by changing the stitching wheels. Custom patterns and special designs are possible by use of custom-made stitching wheels (Figure 15.7).

The sewing table work surface is high-pressure laminate with non-snag, non-glare finish that provides operator comfort and minimizes fatigue. Special accessories such as folders, hemmers, powered top feeders and pullers are available.

Operation

The ultrasonic sewing machine can perform a broad variety of procedures including stitching, tacking, hemming, basting, pleating, slitting, embossing and

Figure 15.7. Additional stitch patterns.

serging. The process can fuse the thermoplastic used to make such fibers as nylon, polyester, polypropylene, modified acrylics, some vinyls, urethane and many synthetic blends with up to 35% natural fiber content.

Ultrasonic sewing is presently being used to make products such as bagging, bandages, blankets, clothing, disposables, draperies, film, filters, packaging, sails, strapping, upholstery and window shades.

Appendixes

Appendix A

Glossary of Ultrasonic Terms

Acoustical Property The ability of material to conduct vibratory energy.

Activity Meter A meter that indicates the power drawn from the generator or power supply.

Amplitude Transformer A half wave-length resonator having a change in cross-sectional area between the input and output surfaces for mechanically altering the amplitude of vibration.

Antinodal Region The area of maximum mechanical amplitude in the direction of sound transmission.

Bonding *See* Plastic Welding.

Booster An amplitude transformer that increases the input amplitude of vibration.

Boss A raised portion in the workpiece, usually of the same material as the workpiece.

Butt Joint Two flat contacting surfaces with no built-in alignment.

Catenoidal Horn A horn having a cross-sectional area following the equation for a catenoid.

Clamping Pressure The pressure exerted by the horn on the workpiece.

Compatibility The relationship between plastics.

Contoured Horn A horn whose frontal surface is machined to conform to the workpiece surface engaged.

Coupler *See* Horn.

Coupling The surface contact between two components.

Coupling Bar A nodal mounted resonant section.

Converter An electro-mechanical assembly in a protective housing.

Deflashing The removal of excess material at the part line of the molded part.

Degating The separation of a part from a molded runner assembly.

Dwell Time The length of time the gauge pressure is maintained on the workpiece after the cessation of ultrasonic energy.

Electro-Mechanical Conversion The conversion of electrical energy into mechanical energy and vice-versa.

Energy Director A projection of plastic material from a part of the workpiece for concentrating the ultrasonic energy along the joint line.

Exponential Horn A horn having a cross-sectional area following an exponential equation.

Far-Field Welding Welding taking place at a distance greater than 1/2 in. from the point of horn contact with the workpiece.

Frontal Surface The horn area that contacts the workpiece.

Full Wave-length Horn A horn one wave-length long in the direction of sound transmitted therethrough.

Fusing *See* Plastic Welding.

Gain The built-in amplitude of a horn or amplitude transformer.

Gang Welder A system having several converters that are sequentially or simultaneously operable.

Generator *See* Power Supply.

Half Wave-Length Horn A horn one-half wave-length long in the direction of sound transmitted therethrough.

Heading *See* Staking.

Hermetic Seal A seal or bond able to withstand positive and negative pressure without leaking.

Hold Time *See* Dwell Time.

Horn Usually a one-half wave-length long resonant bar or metal section transferring energy from the converter to the workpiece, made preferably of aluminum or steel.

Horn Abrasion Material removed from the face of the horn through frictional wear.

Horn Amplitude The peak-to-peak excursion of the horn at its frontal surface (anti-nodal region) during one cycle.

Horn Analyzer An electronic unit for providing such data as horn frequency, horn balance, and Q of the horn.

Horn Change Button A switch for activating the carriage mechanism without ultrasonics, exposing the horn for removal.

Hygroscopic Plastic A plastic that absorbs moisture.

Inch-Pounds Per Second A unit of power 8.85 in.-lbs./sec. equal to one watt.

Insertion The process whereby a metal piece is implanted in plastic.

Interface The area at which two mating parts meet.

Joint Line *See* Interface.

Loading Meter *See* Activity Meter.

Magnetostrictive Effect A means of converting electrical energy into mechanical energy based on the magnetostrictive properties of material.

Mechanical Impedance Transformer An alternate designation for horn.

Mold Release A lubricant that aids in part extraction from the mold.

Motor *See* Converter.

Multi-Probe Welder Assembly units having more than one welding head or converter-horn combination; *see also* Gang Welder.

Near-Field Welding Welding occurring within 1/2 in. from the point of horn-part contact.

Nest A holding fixture for locating and holding workpieces in position for assembly.

Nodal Mounted Pressure Rod A plunger secured to the nodal area of the horn and engaging the workpiece to dampen undesirable vibration manifest at the surface or interface of the workpiece.

Parent Material Strength Strength equal to that of the unwelded joining materials.

Piezoelectric Effect The phenomenon of converting electrical energy into mechanical energy or mechanical energy into electrical energy by use of piezoelectric materials.

Pistol Grip Hand Tool A converter and horn with pistol grip handle for hand-held use.

Plastic Fillers An inexpensive inert material added to the resin to either reduce the cost of plastic or to improve the plastic's physical characteristics.

Post-Cycling Most ultrasonic assembly systems are factory adjusted to trigger the ultrasonics when either a preset pressure is applied on the workpiece or a prescribed time period has elapsed. If this triggering is altered, delaying the factory-set triggering, the system is considered post-cycled or post-triggered.

Power Control A variable control, usually located on the generator, for altering the output power of the generator.

Power Supply The electrical unit that converts low-frequency line power into high-frequency electrical energy.

Pre-Cycling The opposite of Post-Cycling.

Pre-Triggering *See* Pre-Cycling.

Press *See* Stand.

Probe *See* Horn.

Programmer An adjustable control for automatically timing the weld and hold cycles.

Seaming *See* Sewing.

Sewing The continuous welding of thin films or fabrics.

Scan Welder A device for welding flat rigid parts by either transporting the part under one or more stationary horns or scanning the part with movable welding heads.

Slotted Horn A horn that, due to its width, must be slotted to reduce vibrations at 90 degrees to the desired plane "poisson's couplings."

Sound Box A chamber for dampening sound emission to an ambient level.

Spot Welding The process of creating small localized bonds between two parts. parts.

Staking The process of melting and forming the upper portion of a stud or boss in such a manner as to capture and hold another material, usually metal.

Stand The unit that houses the converter and horn in a rigid mounting, allowing it to move up and down either mechanically or pneumatically and applying a predetermined pressure on the workpiece.

Step Horn A horn having a sharp step in cross-sectional area, usually in the nodal region.

Step Joint A joint similar to the butt joint, with the exception of a raised shoulder portion which provides built-in location or alignment.

Swaging A forming of plastic material, which serves to capture and hold another part.

Timer *See* Programmer.

Timing Module The electronic module controlling the pneumatic and electronic functions.

Tool *See* Horn.

Transducer The component in the converter housing that converts electrical energy to mechanical energy.

Tuning the System Matching electrical operating frequency to the mechanical resonant frequency of the transducer assembly and horn.

Ultrasound Vibrations above the audible range of human hearing (18 kHz).

Vacuum Horn A horn with a hole at its frontal surface and a vacuum pump connection on its side for holding small inserts by suction prior to assembly, assuring accurate placement.

Vapor-Proof Converter A converter with an air-tight casing that requires an external air supply for cooling.

Weld Quality Monitor An electronic unit connected to an ultrasonic assembly system for sensing, based on energy measurement, whether an acceptable assembly is achieved.

Appendix B
Federal Conformance

Federal Communications Commission

The Federal Communications Commission has been empowered by Congress to establish rules and regulations concerning the emission of spurious radio frequency radiation from industrial equipment. Ultrasonic equipment can be a serious source of such radiation unless adequate design steps are taken to reduce the radiation to an acceptable and safe level.

According to Part 18 of the FCC Rules & Regulations, all ultrasonic equipment sold after October 1, 1970 must comply with the emission standards established in Part 18 of the Rules & Regulations.

How can you tell whether your equipment meets FCC regulations? If the supplier has submitted the model of equipment for government inspection and it met the requirements, a Type Approval number will have been issued and a label affixed to the equipment indicating the Type Approval number. An alternate method is "Certification." In this procedure, equipment is not submitted for government testing, but the manufacturer certifies to the purchaser that the equipment meets the requirements of the FCC rules and regulations. If the rated output power of the ultrasonic equipment is less than 500 watts, neither Type Approval nor Certification is required, but the equipment still must meet the radiation requirements of the FCC. If a user is in doubt, it is best to get a statement from the supplier to the effect that the equipment in question does meet the requirements of the FCC regulations.

The sale after October 1, 1970 of ultrasonic equipment not meeting the requirements could result in a fine to the seller of up to $1,000 per violation and/or one year in jail.

With respect to the user, continued use of equipment not meeting the FCC requirements, in defiance of an order to comply, will subject the user to the penalties stated above.

Occupational Safety and Health Act (OSHA)

The Williams-Steiger Occupational Safety and Health Act of 1970 (OSHA) contains wide-ranging and stringent provisions for the protection of employees at their place of work. The provisions of this mandatory act, applying to all employers, have a direct impact on the use of ultrasonic welders.

Paragraph 1910.217 of the act states that "mechanical power presses" must have a two-hand trip control

> arranged by design and construction and/or separation to require the use of both hands to trip the press and use a control arrangement requiring concurrent operation of the individual operator's hand controls.

The act further requires that "the control system shall incorporate an anti-repeat feature."

In addition, the moving parts of the stand should be provided with guards and shields to prevent injury to hands if the welder is contacted during operation.

The act requires the employer to meet its provisions. Therefore, installations which fail to comply with the requirements must be upgraded. For older equipment, sold prior to the act, most manufacturers can furnish a conversion kit which should include dual control non-tiedown pushbutton switches to replace either a foot switch or switches which do not include the non-tiedown feature. With the effectiveness of the act, only equipment meeting the requirement of this act can be used by an employer.

The only permissible exceptions to the stated provisions are those installations where the "point of operation" is completely protected, for instance, when sliding enclosures are used or in automated systems where the operator's hands are removed from the working area and/or the operating area during operation is inaccessible to the operator.

The *Federal Register* of Saturday, May 29, 1971, Part II, Vol. 36, No. 105, carries the full text of the act. Several amendments have been issued since. Failure to comply with provisions of OSHA could result in penalties to the employer.

OSHA on Ultrasonics and Noise

Many questions have been asked about the effects of ultrasonics in connection with health and safety regulations relative to noise.

Ultrasonics is inaudible sound. Ultrasonic energy is vibrations at a frequency which is not heard by human beings. The usual cut-off level for the human ear is in the range between 16 and 18 kHz (16,000 to 18,000 cycles per second), depending on the age and sex of the person. With increasing age, the human ear becomes less responsive to higher frequencies and cut-off frequencies between 12 and 16 kHz are common.

Ultrasonic plastics welding equipment operates at an inaudible frequency of 20 kHz or higher. However, in assembling certain plastic parts audible sound (noise) is occasionally generated during the period in which the welder transfers energy to the workpiece. Such sound is produced by the workpiece vibrating at a lower frequency, within the audible range.

Under the Occupational Safety and Health Act (OSHA), certain maximum permissible noise exposures are specified. The exposures are given in Table B-1.

When the daily noise is composed of two or more periods of noise exposure at different levels, their combined effect should be considered, rather than the individual effect of each. If the sum of the following fractions: $C_1/T_1 + C_2/T_2 \ldots C_n/T_n$ exceeds units, then the mixed exposure should be considered to exceed the limit value. C_n indicates the total time of exposure at a specified noise level, and T_n indicates the total time of exposure permitted at that level.

In addition, the act states that "exposure to impulsive or impact noise should not exceed 140 db peak sound pressure level."

Several other considerations should be kept in mind. While 90 dbA is the permissible noise exposure for eight hours, higher noise exposures are permissible for shorter periods of time, provided that the total permissible noise exposure is

Table B.1
Permissible Noise Exposures.

Duration per day, in hours	*Sound level dbA slow*
8	90
6	92
4	95
3	97
2	100
1-1/2	102
1	105
1/2	110
1/4 or less	115

not exceeded. Furthermore, noise measurements must be made at the ear of the worker, and at his normal work position. Hence, if an automatic machine is used and the worker normally wanders away from the machine, the meter must follow the worker.

When making sound-level measurements, it is vitally important that the noise-survey meter has been calibrated. Most meters do not remain calibrated and a proper calibration check prior to taking measurements is a must. Further, the meter must be set to the "A" scale which is a slow-response setting as contrasted with other possible settings usually found on a sound-level meter.

The configuration of the particular part to be welded and the composition of the raw material have a pronounced effect upon noise. In many instances, the noise level is keenly influenced by the hardness or brittleness of the plastics material. A relatively quiet operation can suddenly become noisy when a high percentage of reground styrene plastics, for instance, is mixed into the raw material to reduce material cost. One of the engineering measures that can be taken to reduce the possibility of noise is to pay close attention to the plastics material reaching the molding machine. Generally, virgin and softer plastics material provides a quieter operation than reground and hard, brittle raw material.

Another item that can be responsible for audible noise is the vibration of larger surfaces during welding. The designer of plastic parts intended for ultrasonic assembly should consider the possibility of diaphragming surfaces and, to avoid the possibility of such condition, provide reinforcing ribs to reduce vibration. Alternatively, it may be advantageous to divide continuing surfaces into smaller sections using steps, radii, stepped wall thicknesses, etc. During welding, vibrating surfaces may need to be mechanically supported to dampen large-amplitude vibration.

If the exposures as previously described are not met or the measures listed are unsuccessful, noise can be reduced by providing partial or complete enclosures. The enclosures, most suitably, are composed of wood or metal-backed, sound-absorbing foam. The foam faces the welding horn, and the wood or metal surface faces the area of the workers. Instruction manuals supplied with welding equipment list several manufacturers of sound-absorbing material. Application Laboratory generally makes a notation on the laboratory report if an operation is noisy and noise abatement equipment is recommended, and such information is conveyed to the customer. If noise abatement enclosures are specified by the customer, these can be provided as part of the welding equipment.

Other measures include administrative controls which, as defined by the government, involve rotating employees to limit the exposure of each one. Finally, operating personnal may have to be provided with personal protective equipment such as commercially available earplugs and earmuffs. Earplugs must be fitted for effectiveness. (It should be noted that ordinary cotton stuffed into the ear is not acceptable under OSHA.) Sources for such protective equipment are listed in the operating manual supplied by Branson.

The following publications may be found helpful: "Primer of Plant Noise Measurement," General Radio Company, West Concord, Massachusetts 01781; "Guidelines to the Department of Labor's Occupational Noise Standards," Bulletin 334, U. S. Department of Labor Occupational Safety and Health Administration, Washington, D. C. 20210. Similar publications are available from several manufacturers of noise measuring equipment.

References

Branson Sonic Power Company, Danbury, Connecticut
Ultrasonic Seal Company, Bromall, Pennsylvania
Ultrasonic Systems, Inc., Farmingdale, New York
Macrosonics Division, Paramus, New Jersey
Gulton Industries, Inc., Schiller Park, Illinois
Kleer-Vu Industries, Inc., Ardmore, Pennsylvania
Sonabond Corporation, West Chester, Pennsylvania
Aeroprojects, Inc., West Chester, Pennsylvania
Cavitron Ultrasonics, Inc., Long Island City, New York
Blackstone Ultrasonics, Sheffield, Pennsylvania
Society of Electronic Engineers, New York, New York
International Ultrasonics, Inc., Gilalton, Indiana
Unitek Weldmatic, Inc., Monrovia, California
American Welding Society, New York, New York
Ultra-Clean, Inc., Orange, Connecticut
MacGregor Associates, Inc., Liburn, Georgia
Grover Equipment Company, Louisville, Kentucky
The Crump Company, Denver, Colorado

Index